PRINCIPLES OF PROCESS RESEARCH AND CHEMICAL DEVELOPMENT IN THE PHARMACEUTICAL INDUSTRY

ABOUT THE AUTHOR

Oljan Repič was awarded his B.S. degree in chemistry from the Massachusetts Institute of Technology and a Ph.D. degree in organic chemistry from Harvard University under the direction of Professor R. B. Woodward. As an undergraduate student, he was a teaching assistant at M.I.T. and worked in the laboratory of Professor K. B. Sharpless. As a graduate student, he was a teaching fellow and a mass spectrograph operator for Professor Woodward's group. During this period, he also translated R. B. Woodward and R. Hoffmann's book *The Conservation of Orbital Symmetry* into the Slovenian language. Dr. Repič stayed with Professor Woodward for another year as a postdoctoral fellow to pursue the first synthesis of dodecahedrane. Thereafter, he joined Sandoz Pharmaceuticals Corp. in East Hanover, New Jersey, as a scientist in chemical research and development where he has been given increasing responsibilities and is today the director of process research and development. His group has designed the chemical process for the marketed drug Lescol and many other drugs in clinical trials. Dr. Repič has co-authored more than 45 papers, holds five patents, and has given more than 20 invited lectures around the world on topics described in this book. He is a member of the Phi Beta Kappa and Phi Lambda Upsilon honor societies, the American Chemical Society, and the Drug Information Association.

PRINCIPLES OF PROCESS RESEARCH AND CHEMICAL DEVELOPMENT IN THE PHARMACEUTICAL INDUSTRY

Oljan Repič, Ph.D.

Sandoz Pharmaceuticals Corporation
East Hanover, New Jersey

A Wiley-Interscience Publication

JOHN WILEY & SONS, INC.

New York / Chichester / Weinheim / Brisbane / Singapore / Toronto

This book is printed on acid-free paper.⊚

To order books or for customer service please, call 1(800)-CALL-WILEY (225-5945).

Library of Congress Cataloging-in-Publication Data

Repič, Oljan, 1948–
    Principles of process research and chemical development in the pharmaceutical industry / Oljan Repič.
        p.   cm.
    "A Wiley-Interscience publication."
    Includes bibliographical references and index.
    ISBN 0-471-16516-6 (cloth : alk. paper)
    1. Pharmaceutical chemistry.   2. Chemical process control.
3. Organic chemistry I. Title.
    [DNLM: 1. Technology, Pharmaceutical—methods. 2. Chemistry,
Pharmaceutical—methods.   QV 778 R425p 1998]
    RS403.R46   1998
    615'. 19—dc21
    DNLM/DLC
    for Library of Congress   97-13904

To my father,
Riko Repič, Sc.D.
Professor Emeritus, University of Ljubljana

CONTENTS

FOREWORD

DEVELOPMENT PERSPECTIVES
RESPONSIBLE PROCESS DEVELOPMENT

The following chapters have been written in testimony to the often unsung efforts of process research and development. After all, the process to put a new drug on the market only begins with discovery. We have to put the genie in the bottle. For those of you who wish to understand how this can be done, this book is for you.

Over the years I have identified some key areas that I believe to be of utmost importance to responsible process development. This process entails balancing the aspects of speed, economy, ecology, safety, regulation, teamsmanship, and pride throughout the development cycle. At times, each of these areas will hold greater significance; however, seasoned scientists and managers will strive to integrate as many of these areas as early as practicable.

PRIDE

Although listed last above it is prudent to begin with pride. It is most important that process chemists develop a sense of ownership and responsibility for their work. After all, discovery has been accomplished elsewhere. The best process chemists exercise creativity in every stage of their development efforts and take great satisfaction in showing off their work. Whether it be the brevity of their synthesis compared to the original, the cost reduction, yield improvement, development of new reagents for synthesis, or the breakthrough understanding of reactive processes; recognition of this work is what the process chemist thrives on. Thus it is especially important that process chemists be allowed and encouraged to patent their work and to publish in peer-reviewed journals.

BEGIN WITH THE END IN MIND

One approach to process development is to begin with the end in mind, but to keep in mind that no synthesis can ever be fully developed. Review the current synthesis and be ever vigilant to devise a better route. Generally, the fewer the steps, the cheaper and faster the synthesis (there are some major exceptions, however, and examples will be provided later). Do not believe everything you read, especially the claims for reactions that "do not work." And above all, do not be afraid to experiment. A simple calculation reflects the benefit of responsible process development. A typical optimized drug cost of $<\$1,000/kg$ that could realistically represent a 90% reduction of an underdeveloped process costing $\$10,000/kg$ can save $\$9,000/kg$. At a modest 1,000 kg/year production rate, this represents a yearly savings of $\$9$ million. This also allows preclinical chemists to bring forward molecules of ever greater complexity, comfortable with the knowledge that process research and development will work their miracles.

SPEED

In my opinion, speed is often used as an excuse for haste and an unwillingness to explore. Quite often a few good ideas and some quick experiments will dramatically improve a synthesis. Dedicated and concurrent development, however, yields exceptional results. A few weeks or months of effort early in the development cycle will pay off handsomely in shorter and more productive laboratory and plant runs. Each time a delivery synthesis is run, it should become better. Be forever on the lookout for ways to streamline the synthesis as you develop it into a process. One technique used to build in flexibility is to concentrate on the last steps first. Seek to identify and develop the final step of a synthesis from an isolated and characterizable intermediate as quickly as possible. Develop a robust and forgiving crystallization that rejects impurities and then lock these steps in. Subsequent and different front end chemistries will then more easily meet the capabilities of these last steps and cause much less concern to the regulatory bodies.

FOREST FOR THE TREES

There is an old saying that sometimes one cannot see the forest for the trees. This is a trap into which many process chemists fall. When testing a new synthetic route it is imperative that the process produces acceptable material. Often a chemist will spend countless hours optimizing a particular step only to find out that the subsequent and/or anticipated step(s) fail(s). When you've got an idea that looks promising, pursue it through to the end to determine if you can make the desired product of acceptable quality, then go back and clean up the details.

THINK STEP INTEGRATION

How well does the process flow? Ask yourself these questions: Is the product from the preceding step, including solvent(s), compatible with the next step? Must the product be isolated or can it be used in the workup solvent? Are there too many unnecessary isolations? Have I used azeotropes (binary and/or tertiary) to remove water and incompatible solvents instead of solid drying agents and crystallizations and oven drying?

ECONOMY

Do you realize that what you discard during a process often costs more than what you make? The chemist fresh from graduate school will often run reactions at minimal concentrations (100 mg substrate in 100 mL of solvent is not atypical). This practice is extremely costly. Experience has taught us that many reactions run well at concentrations exceeding 100 mg/mL or 100 g/L. Bimolecular reactions under high concentration run faster, are over sooner, and produce less solvent waste per unit product. Be aware that some reactions can even run without solvent(s); however, solvents are often needed to remove products from reactor vessels.

SAFETY

Safety is one area that cannot be stressed enough. All responsible pharmaceutical firms have invested heavily in process safety evaluation. The intrinsic energetics of molecular structures must be known and dealt with accordingly. It is the unknown that can kill. Particularly important is to observe the heat effects from high-concentration and biphasic reactions. With such little solvent ballast to soak up the heats of reaction, high-concentration reactions must be considered to run adiabatically. Fortunately, these reactions can often be throttled through addition control. Biphasic reactions are extremely sensitive to mixing and easy to overcharge if one is relying on addition control. Be particularly vigilant for autocatalytic reactions, as these can often lead to dangerous situations.

ECOLOGY

Think recycle, even if you do not plan to use it. If your product goes commercial your colleagues in manufacturing will thank you. Invariably most processes will use solvents. Judicious choice thereof will allow for easy recovery, and reuse can dramatically affect the cost of a process. Every piloted synthesis should include a method for rendering spent reagents in as least harmful or toxic form as possible. This has a dramatic effect on disposal costs and mitigates the potential for environmental exposure downstream.

REGULATION

Perhaps no single event in the history of pharmaceutical development has engendered such passionate debate, concern, reaction, interpretation, and investment as good manufacturing practices (GMP). Because of this, it has been described as an ideology, a religion, and a philosophy. Since the interpretation and definition of GMP is continually changing it is now more commonly referred to as current good manufacturing practices (CGMP). I believe that the basis for GMP lies in good science, ethics, and documentation and differs little, in principle, from how a well-trained scientist should maintain his or her laboratory journal. The basic teachings are (1) traceability of raw materials (solvents, reagents, etc.), (2) cleanliness (what is being done to avoid cross-contamination), and (3) quality of product (what evidence do you have that what you think you have is indeed what you have?).

Certainly GMP came about for good reasons. This cannot be discussed here; however, it is acknowledged that the public at large needed protection, and regulatory agencies worldwide endorsed the concept. Although these regulations are quite clear for production, it is indeed a challenge to apply these to the experimental nature of development (e.g., surprise, flexibility, failure, and rework). At this writing well-intentioned and pragmatic regulatory lobbyists are trying to bring some welcome interpretation to the confusion. Simply put, at the beginning of development it is the product that defines the process and at the end of development it is the process that defines the product. It is only a matter of definition, so stay tuned.

Thus said, I wish you successful development. Learn from others, and above all learn from your mistakes.

THOMAS J. BLACKLOCK, PH.D.

Vice President
Novartis Pharmaceuticals Corporation

PREFACE

Chemical development, unlike most other areas of chemistry, is not taught in schools or in textbooks. Chemical development scientists are made on the job, only after several years of experience working in a laboratory next to a production plant, who is the customer.

With this book I would like to fill this gap: It will help students decide whether to choose an industrial career, it will give a head start to new development chemists (who will have the answers at the start instead of wondering for a few years what it is all about), it will remind established development chemists what the goals and the focus really should be, and it will convince academic chemists that process research and development is the most innovative and rewarding area of industrial organic chemistry. With a few exceptions, the message will be conveyed with examples of recent organic chemistry, in many figures and tables.

Most graduate students in organic chemistry have a list of desirable careers, and it usually lists this order of priorities: academic research (a university professor), industrial research (a medicinal chemist), industrial development (a development chemist). Research always seems to come before development because research is perceived as more glamorous. I also had to be convinced that chemical development is a desirable career; I must give most credit for this conversion to my colleague, Dr. Jürgen Martens, who argued approximately as follows.

The odds of discovering a marketable drug within one life as a medicinal chemist are very small. The career of a development chemist is much more satisfying because visible results come much more often. For example, since one works only on advanced projects, the chances of one of them reaching the market are much better, and it is most gratifying to see your own chemical process produce a lifesaving drug substance.

The other misconception is that chemical development is not very creative. This could not be further from the truth, as this book will show. The medicinal chemist may synthesize a drug substance in 10 steps, with a chromatography in each step, with a total yield of 3%. The development chemist would prepare the same molecule in four steps, in 65% total yield, without any chromatography, at one-tenth of the price. This requires much creativity, design of new syntheses, and an intimate understanding of the chemistry involved. Anyhow, the medicinal chemists are becoming more biochemists as they seek to understand biochemical mechanisms of diseases; the only department that still studies organic synthesis for its own sake is chemical development. In fact, graduate school research in organic synthesis resembles chemical development much more than it resembles medicinal chemistry: Given a target molecule, one must design the synthesis and discover and *develop* all reaction conditions to obtain a reasonable yield of the target molecule.

Chemical development is complex, exciting, varied, interesting, and always challenging. One sees all the active projects coming through the research pipeline; one works on all different chemical classes of compounds; and the results are quick, visible, and satisfying. I never regretted the choice of this career path. In this book, I attempted to convey these observations and feelings.

Chapter 1 describes the basic principles of chemical development and is a mission statement that guides the reader through the rest of the book. The qualities of an ideal process are explained and illustrated with two real-life chemical development examples. We learn that it is the process chemist's responsibility to ensure that drug substances meet purity specifications. If impurities exceed the limits, the process chemist must minimize these by-products.

Trouble-shooting impurities is the most intriguing activity in chemical development because this task is intellectually challenging, like a detective game; to stimulate readers' interest, the topic is introduced in Chapter 2. An unknown impurity appears in the final product that causes it to fail specifications. What now? The strategy is to (*1*) isolate the by-product by prep high-pressure liquid chromatography (HPLC); (*2*) determine its structure by NMR and mass spectroscopy (MS); (*3*) propose a mechanism for its formation; (*4*) identify the synthetic step in which it forms; and (*5*) change the reaction conditions, using the mechanistic hypothesis, to minimize the formation of the undesirable by-product. Twenty examples are described. The most satisfying moment for development chemists is the introduction of their new drug into the market, when years of hard work are rewarded by the knowledge that their process is producing literally tons of a marketed drug. I had the fortune of being associated with one such product, Lescol, a lifesaving drug that lowers serum cholesterol, from its discovery in the research laboratory through its introduction to the market. Chemical development was involved throughout this period of about 10 years. During this time, innovative chemistry was required to design the ideal process.

Chapter 3 describes this 10-year effort with one drug, a racemic compound. In the future, however, more drug substances will be enantiopure. The reasons for this are discussed, and three methods of synthesizing enantiopure compounds are presented and exemplified in the following three chapters: the chiral pool strategy, asymmetric synthesis, and resolutions. Guidelines of a good asymmetric synthesis are given.

Most of Chapter 4 is devoted to resolutions, and various methods of resolution are illustrated: by preferential crystallization, by chromatography, kinetic resolution, via diastereomers, resolutions with 100% yield, and spontaneous resolutions. Criteria for an ideal resolving agent are revealed.

Asymmetric synthesis is emphasized in Chapter 5 and is illustrated by three different syntheses of each enantiomer of one potential drug, SDZ 62-834. This compound is a most interesting example, because both pure enantiomers were needed. The enantiomers differ not only in activity but also in indication: One was of interest in the area of oncology, the other against multiple sclerosis.

The chiral pool approach is demonstrated in Chapter 6. This strategy looks to nature to provide enantiomerically pure building blocks from which the optically active drugs can be constructed. Chapter 6 is the climax to which the preceding discussions were leading: how to synthesize the cholesterol-lowering agent that was introduced as a racemate in Chapter 3 in an enantiopure form. Three syntheses are described: the first one starting from D-glucose, the second from L-malic acid, and the third one from L-mandelic acid. The first two use the chiral pool strategies, the third one is, strictly speaking, an asymmetric synthesis (the difference being whether the chiron ends as part of the product or whether it is a recyclable reagent).

The moral of Chapter 6 is that asymmetric syntheses are usually more efficient than natural-product strategies. For example, an asymmetric synthesis can be as short as a racemic synthesis (just replace one reagent with a transient chiron). The second lesson is that reducing the number of steps in a synthesis—by designing a shorter synthesis—is the most effective method for reducing the cost of a process.

To study the fate and distribution of a drug in the body or in the environment, drug substances for these experiments are often labeled with isotopes. Synthetically, the easiest strategy for introducing a label is to replace a hydrogen with deuterium or tritium; on the other hand, hydrogen labels are also lost most easily as many metabolic reactions in the body remove hydrogen atoms from molecules as water. It is, therefore, important to find a biologically stable position for the label; any loss of the label could lead to a false quantification of the drug in biological samples. Therefore, carbon isotopes are often preferred; however, such syntheses that incorporate carbon isotopes into the drug substance are usually longer, because the carbon label, by definition, must be built into the skeleton of the molecule. Many examples of syntheses that label drug substances with ^{2}H, ^{3}H, ^{13}C, or ^{14}C illustrate these principles in Chapter 7. This chemistry radiates excitement, not only β-particles!

The pharmaceutical industry is among the most regulated industries in the United States. The story of chemical development is not complete without Chapter 8 on government regulations. Three areas that affect chemical development are the development report, process validation, and good manufacturing practices. These three regulations are explained but made digestible by presenting real examples and data.

The book concludes with Chapter 9 on speculations about the future of chemical development, emphasizing the promise of monoclonal antibodies as custom-designed catalysts.

I thank Dr. Denise Pasternak, associate editor at John Wiley & Sons, who gave me the confidence to undertake the task of writing this book. I am grateful to my colleagues Dr. Thomas J. Blacklock, Dr. Prasad Kapa, Dr. Mahavir Prashad, Dr. Orin Tempkin, and Dr. David Xu for reading and commenting on the manuscript. Christine Foster drew all the chemical schemes, Maurice Castelbuono drew the other art, and Divya Sahai performed many literature searches. Thanks to all the past and present members of my research group for the excellent chemistry described in this book; their names appear in the references after each chapter. I must also give credit to Professor D. S. Kemp, the best teacher I ever had the privilege to listen to, who could explain the mysteries and convey the magic of organic chemistry like none else.

Thanks to Susan, my truest and sweetest companion, and to Oriana and Adrian who had to eat dinners without me but cheered me along anyway.

OLJAN REPIČ

PRINCIPLES OF PROCESS RESEARCH AND CHEMICAL DEVELOPMENT IN THE PHARMACEUTICAL INDUSTRY

INTRODUCTION

A popular cliché in philosophy says that science is pure analysis or reductionism, like taking the rainbow to pieces, and art is pure synthesis, putting the rainbow together. This is not so.

—J. Bronowski, *The Ascent of Man* (1975)

Let me begin with an anecdote that I have every reason to believe actually happened:

At one time in a well-known chemical firm, a chemist synthesized a fine-looking compound. He scraped some of it out of his reaction flask and did some tests. Lo and behold, he saw it was good. He communicated with his supervisor and so on. Before long, it was decided to run a fairly large batch in the pilot plant. The chemist dutifully wrote up the method of creating this wonderful material and passed it on to the plant. About a week later he got a call from someone engaged in making this new material. This someone was curious about the method by which this material was handled after preparation. The chemist replied that he merely scraped the stuff out of the flask with a spatula. The voice on the telephone said: "Well, get a great big spatula and come right over. We've got 2000 pounds of it stuck in this reactor."

This story dramatizes the need for process development. Let me make the point in one other way. Most people, when they hear the term chemical development say, "Oh, you scale up syntheses." If I leave you with only one message after reading this book, I would like it to be the message that scaling up is only a small part of process development.

We need process development precisely because most research procedures cannot be scaled up. Medicinal chemists occasionally say, "If we can make 1 g, how come you have so much trouble producing 1 kg?" To illustrate why we cannot simply multiply the research procedure by a factor of 100 or 1000, let us look at

Table I.1, which shows, in part, what we variously call *synchronization* or *cost estimate* because it shows the amount of each reagent needed to make 1 kg of the drug substance (column 3) and the cost each reagent contributes to the cost of 1 kg of drug substance (column 5). If one simply multiplies the research procedure (column 2) by a factor of about 200 here, one obtains some bizarre numbers (column 3). To make 1 kg of this drug substance, we need almost 900 L of diethyl ether, about 100 kg of starting material, >100 kg of mercuric acetate, which, by the way, produces 100 kg of elemental mercury as a waste product—all this for each kilogram of drug substance, which would end costing $48,000/kg. These are large numbers by any standards.

We go through this exercise with every project, not to slander the research procedure but to identify weak points that require process development, either in the area of raw material cost ($48,000), volumes (100 L/kg), safety (diethyl ether), or ecology (mercury). By *research procedure,* I mean an undeveloped procedure; it may come from discovery or from our own process research and development (R&D) laboratories.

So then, what is the function of chemical development? Chemical development is positioned between research and production and is involved with drug development of each project throughout this period, which is 5–10 years. Chemical development has two major functions. One of them is to produce all the drug substance needed by various other departments throughout these 5–10 years. The medicinal chemists typically prepare only a few grams of material that is used for biological screening. In the first phase of development, the preparation laboratory prepares about 1 kg of drug substance. It is used for short-term toxicology, for stability testing, by analytical research, and by pharmaceutical development. For the next stage

Table I.1. Cost Estimate and Synchronization of a Chemical Process

Ingredient	g/Reaction	kg/kg of Product	$/kg	$/kg of Product
Step 2				
D-Glucal	545	97.16	193.00	18,751.81
Sodium	2	0.36	22.20	7.92
Mercuric acetate	651	116.06	22.00	2,553.25
Methanol	902	160.80	0.27	43.42
Celite	70	12.48	4.15	51.79
Intermediate 2	608	108.39	197.51	21,408.19
Step 25				
Intermediate 24	0.23	2.83	16,341.08	46,299.74
Tetrahydrofuran	3	35.71	2.27	81.07
Acetic acid	0.12	11.43	0.51	0.73
$n\mathrm{Bu_4NF}$	0.41	4.88	178.07	869.15
Diethyl ether	75	892.86	0.99	883.93
Drug substance	0.08	1.00	48,134.62	48,134.62

of development, about 10 kg of material is prepared in a kilogram plant. This material is used for long-term toxicology and by drug delivery systems. The latter consume much of the material, as researchers perform physical tests, for example, on how to form tablets. This material may also be used for the first clinical studies that determine the safety of the drug. If the project advances, about 100 kg more of the drug substance is needed. This is prepared in the pilot plant and is used for longer-term (phases II and III) clinical tests that establish the safety and efficacy of the new drug. When the product reaches the market, several tons of drug substance may be needed every year. These numbers are just order-of-magnitude figures and depend on the drug indication, dose, and market size.

The second major function of chemical development is process development. This activity is parallel to the preparation activities and is performed first on a laboratory scale. The input to chemical development is the research synthetic procedure; the output is drug substances and commercial plant processes. In addition, reports are written that are incorporated into Investigational New Drug (IND) and New Drug Application (NDA) documents filed with various governments to obtain approvals for clinical testing and for marketing, respectively.

The objective of process development is to make kilogram-scale production of drug substances possible and economical. These two objectives do not have to be reached simultaneously: Usually, we first make the production possible; and only in the later phases of development, when the scale becomes significant and certainly before marketing, do we have to lower the cost of the process. In other words, the function of process R&D is to transform a research synthetic procedure into a plant process by performing the necessary laboratory experiments to achieve the goal of an ideal process. In Chapter 1, we will see what makes a process ideal.

1

THE IDEAL*

What qualities should an ideal plant process have? The ideal process should

- Be safe: safety is always the number one priority.
- Be ecologically sound.
- Be reproducible both on repetitive runs and on scale up.
- Fit the plant physically (in the development phase we usually fit the chemistry to the plant and not the plant to the chemistry).
- Produce product of high quality that passes predetermined specifications.
- Be economical.

As the word ideal implies, there is no ideal process. The best we can do is to approach the ideal, and our success is measured by how closely to it we can come. Clearly, some of these qualities oppose each other; for example, typically the higher the desired purity, the lower the yield will be and the higher the cost. Compromises must be made.

1.1. SAFETY

No process can be made 100% safe; however, one can identify hazards and then, if the hazard is severe, one avoids such a chemical or process, or if the hazard is moderate, one takes precautions that minimize the risk. What are some safety hazards? They can be divided into four categories: thermal instability, toxicity, flammability, and explosiveness.

*Based on a lecture I gave at the Pharmaceutical Manufacturers Association meeting, Indianapolis, Ind., Apr. 14, 1986.

1.1.1. Thermal Instability

Our in-house safety laboratory measures the decomposition temperature of all reagents, intermediates, solvents, distillation residues and evaporation residues, and any exotherms associated with the decomposition, so that one stays well below these temperatures in the process. We also measure heats of reactions (calorimetry) to ensure adequate cooling capacity of the reactor before scaling up a reaction. This minimizes the risk of runaway reactions.

1.1.2. Toxicity

By *toxicity* one means, for example, LD_{50} (projected lethal dose that would kill half the experimental animals) and carcinogenicity. For known compounds, one looks up such data in manuals, such as Sax's *Dangerous Properties of Industrial Materials*. For new compounds, we perform these tests in-house, either the LD_{50} or the Ames test, respectively.

There is no hard rule on what is too toxic to be acceptable in the plant (1), but the rule of thumb may be the following (Table 1.1): We would avoid reagents or intermediates that have an LD_{50} <100 mg/kg, which corresponds to a lethal human dose of a teaspoon. But we do work with many solvents that have an LD_{50} of 200 mg/kg. We do not use solvents that are carcinogens, e.g., chloroform, carbon tetrachloride, dioxane, benzene, and hexamethylphosphoric triamide.

1.1.3. Flammability

Another safety hazard is flammability. It is a concern mainly with solvents that are volatile. Since most solvents are well known, one can find the following data in manuals (Table 1.2): ignition temperature (at which the compound ignites), flash point (at which the vapors ignite in the presence of an ignition source like a spark or a flame), vapor pressure (measure of solvent volatility), vapor density (do the vapors rise or creep along the floor?), and the mixture of solvent and air that is flammable. The lower and the wider the latter value, the more hazardous the solvent, since the more likely one encounters the conditions for flammability in the plant.

Table 1.1. Toxicity of Chemicals[a]

LD_{50} (mg/kg)	Toxic	Human Lethal Dose
<1	Dangerously	A taste
1–50	Seriously	1 teaspoon
50–500	Highly	1 ounce
500–5,000	Moderately	1 pint
5,000–15,000	Slightly	1 quart
>15,000	Low	1 quart

[a]Reprinted with permission from Ref. 2.

Table 1.2. Flammability of Solvents

Solvent	Mixture with Air, %	Flash Point, °C	Ignition Temperature, °C
Dichloromethane	12–19	—	615
Acetone	3–13	–18	465
Carbon disulfide	13–50	–30	90
Ethyl acetate	2–11	–4	427
Ethanol	3–19	13	365
Diethyl ether	2–36	–45	160
Heptane	1–7	–4	225
Hexane	1–7	–22	260

Again, there is no hard line between what is safe and what is hazardous, but a rule of thumb may be that we would not work with solvents that have a flash point of less than −18°C; this rule would, for example, exclude diethyl ether and carbon disulfide, but include acetone. Carbon disulfide is also hazardous because of its low ignition point: It can ignite on a steam pipe. Similarly, diethyl ether can ignite on a hot plate or on a hot heating mantle. It is especially hazardous because it is heavier than air and travels along the floor to find ignition sources farther away. Diethyl ether is also hazardous because of its low and wide mixture with air range that is flammable (2 → 36%). Professor Büchi of M.I.T., the story goes, once extinguished his cigarette in a beaker full of ether to calm a student's fear of the lit cigarette. This anecdote proves only that the mixture of ether and air must be lower than 36% to ignite not that ether is a safe solvent.

With moderately hazardous solvents (flash point above −18°C), we take precautions such as grounding of all drums and reactors to prevent buildup of electrostatic electricity and excluding all other sources of ignition. More hazardous solvents (toxic or flash point < −18°C) must be replaced. For example, chlorinated solvents (carbon tetrachloride, chloroform, dichloromethane), most of which are carcinogenic, are replaced with ethyl acetate; pentane (flammable) and hexane (forms electrostatic charge) are replaced with heptane; benzene (carcinogenic) with toluene or xylene; diethyl ether (flammable) with *t*-butyl methyl ether; dioxane (carcinogen) with tetrahydrofuran; hexamethylphosphoric triamide (carcinogenic) with N,N'-dimethylpropyleneurea; and so forth. *t*-Butyl methyl ether is an interesting solvent: It is cheap, relatively safe, does not form peroxides like other ethers, and is not toxic. It has a wide range of uses, from gasoline additive to medicine (it is used to dissolve gallstones, which are made of cholesterol, by direct injection into the gallbladder). However, rats that ingested high doses of this solvent (1000 mg/kg) did show an increased incidence of cancers.

We have exemplified the first strategy of a process chemist: Replace unsafe (toxic or flammable) solvents and reagents with safer ones. It is the chemist's task to determine, however, if these changes adversely affect the reaction chemistry. Although by definition solvents do not participate in reactions, they can dramatically influence the rate and change the course of some reactions (see examples in Chapter 2) (3).

1.1.4. Explosiveness

Another safety issue is explosiveness of compounds. This factor is evaluated in our in-house safety laboratory with tests such as the dust explosion test (a spark is used to ignite a cloud of compound dust) and the hammer test (in which a weight is dropped on a sample of the test compound, and any associated sound or light is recorded, which would suggest an explosion). A good way of predicting explosiveness of compounds is to look at their chemical structure.

Compounds containing weak bonds (e.g., between heteroatoms as in peroxides, hydrazines, halamines, and hydroxylamines) and compounds that can eliminate small, stable molecules like N_2, O_2, NO, or NO_2 (e.g., diazonium, ozonides, nitroso, and nitro compounds) are potentially explosive, because such decompositions, if allowed to proceed, would be exothermic (in the first case because the energy of the starting material is high, in the second case because the energy of the products is low).

1.2. ECOLOGICALLY SOUND

An ideal process must be environmentally acceptable. Nowadays ecology is a number two priority, right after safety. The goal can be achieved by:

· Recycling solvents and reagents (also good for economic reasons).
· Minimizing the amount of solvents and reagents (also good for economics).
· Avoiding toxic solvents and reagents (also for safety reasons).
· Avoiding solvents with low boiling points (as they escape into the atmosphere on distillation).
· Avoiding mixed solvents (for easier distillation, recycling, or disposal).
· Using catalysts, which by definition are used only in small amounts.
· Using "clean" chemistry, e.g., electrochemistry and photochemistry, which by definition use no reagents only electrons or photons, respectively (4,5).
· Using water as a solvent.
· Using unconventional energy sources (ultrasound, microwaves).

1 3. REPRODUCIBLE

A process must be reproducible on repetition and on scale up. Chemical processes are quite reproducible if all critical parameters are kept constant or, more practically, are kept within specified limits. This philosophy is the basis of process validation (see Chapter 8). The difficulty, of course, is to be able to identify all critical variables. For this reason, process reproducibility is not tested by statistical analysis (i.e., by running the process 10 times). Instead, reproducibility is built into a process by process development (i.e., by identifying and defining all critical process parameters). So processes developed on a laboratory scale should scale up predictably in the plant; this belief is the premise and justification for process development laboratories.

Having said that, however, of all the qualities of an ideal process, the reproducibility on scale up is most difficult to check on a laboratory scale, for obvious reasons. As much as one tries to define all parameters and as much as one tries to simulate plant conditions in the laboratory, some physical parameters will be different in the plant: stirring efficiency, surface area: volume ratio of a reactor compared to a lab flask, the rate of heat transfer, and the temperature gradient between the center of the reactor and the walls, for example, are all different on scale up. The difference is especially noticeable with heterogeneous mixtures or with gasses when stirring efficiency is critical. On the other hand, the plant results are not always worse; often crystallizations are cleaner, with higher yields and purity than in the laboratory.

It is for these reasons that we have development plants and do not go from the laboratory directly into production plants. This thought brings us to the next quality of an ideal process: It must fit the plant.

1.4. PLANT FIT

A process must fit physically into a specific plant or reactor. In going from a research procedure to a plant process, several physical parameters usually need to be adjusted.

Reactants. Reactants must flow in and out of reactors; one cannot scrape solids from the walls of the reactor or turn it upside down as one does with a laboratory flask.

Volumes. Volumes of solvents are usually too large in research procedures and must be reduced to a solvent: solute ratio of 5:1 or less; extreme total volume changes should be avoided so that one is not below the stirrer level at the beginning of the reaction and climbing up the condenser at the end of the reaction or workup.

Reaction Temperatures. Reaction temperatures must be adjusted so that the plant can achieve them in a specific reactor. For example, exothermic reactions are difficult to keep near the limit of the coolant in the reactor (e.g., at $-20°C$ with glycol or at $15°C$ with water cooling). Similarly, it is difficult to maintain nonexothermic reactions far away from the temperature of the coolant, for example, at $5°C$. The heat source in the laboratory is usually a heating mantle. For the medium-scale experiments performed in process laboratories (up to 22-L flasks), heating mantles are safer than oil baths. The disadvantage of heating mantles, however, is that it is more difficult to measure the outside temperature of the flask and that they overheat. Recording both the inside and outside temperature of a reactor (e.g., with a two-pen recorder) is most desirable, among other reasons because the difference between the two indicates an exotherm.

Process Timing. Most operations, like additions or distillations, take longer on a plant scale (but not reactions, because their rates depend on concentration not scale); this time must be duplicated in the laboratory. For example, when the research procedure asks to "add the reagent at once in one portion," this operation has to be translated into "add quickly over 15 min" for the plant. Similarly, all reactions and workup must fit into the workday of the plant, especially if there are only one or two shifts.

Distillations. The temperature and vacuum must be adjusted to plant capabilities. To avoid overheating and because dry solids cannot be removed from the reactor, one should avoid evaporating solutions to dryness. Oily products tend to retain as much as 50% of solvent, because solvent cannot be removed further under the moderate heat and vacuum conditions encountered in reactors. It can be removed by chasing it out with the solvent from the next step of the synthesis. Sometimes the excess solvent is compatible with the next operation, in which case it can stay, and it even helps the draining of the reactor. For similar reasons, the use of rotary evaporators is discouraged, since they do not simulate plant conditions. Solvents should instead be distilled from the reaction vessel to mimic the long distillation path encountered in a plant reactor, or a wiped-film evaporator can be used.

Crystallizations. Cooling rates must be specified so that suspensions are stirrable and not too thick and so that correct crystalline forms are obtained. Stirring rates and seeding are also variables that must be defined.

Filtrations. Hot filtrations should be avoided or performed at least 15°C above the crystallization temperature to prevent crystallization during filtration. Filtration times should be optimized and the plugging of the filter should be avoided by eliminating pastes or fine crystals.

Extractions. Emulsions must be avoided and separation times and the number of extractions minimized. Countercurrent extraction is an efficient alternative. Avoid drying agents for solvents (e.g., sodium sulfate), both before use and after workup, as the plant dislikes the many operations it entails. Instead, use brine to dry organic solvents by extraction or use toluene to dry the product by azeotropic distillation.

Most of the physical parameters I have described can be studied and defined in the laboratory.

1.5. SPECIFICATIONS

An important job of a process development chemist is to ensure that the final product meets or exceeds quality specifications. With drug substances, this is an extensive list of perhaps two dozen properties that must be within predetermined limits; for example, melting point, color of solution, loss on drying (implies solvents), residue on ignition (implies inorganic impurities such as silica gel or glass), particle size, polymorphism, solvates (these three will affect the bioavailability of a drug substance), chemical purity, stereochemical purity, by-product content (identity and amount), heavy-metal content, and pH. Table 1.3 illustrates some specifications for solvents and for heavy metals.

These limits are proportional to the toxicity of the impurities; note that the specification for carcinogens is not detectable, which in practice means that they cannot be used in the last few steps of the synthesis, because their removal to nondetectable levels would be difficult. The specifications may have to be changed if the dose of the drug is especially large, in which case permitted daily exposure (PDE) in milligrams per day may be a better unit for specifications than percent (6).

1.6. ECONOMICAL

A process can be made less expensive in many ways, for example,

- By reducing the number of synthetic steps, which can be done by redesigning the synthetic strategy.
- By improving reaction yields, which can be accomplished by optimizing reaction and purification conditions.
- By lowering the cost of raw materials and reagents, which is done by changing the synthetic strategy, by making the raw material in-house, or by finding a less expensive source (e.g., by competitive bidding and with the help of the purchasing department).
- By running larger batches, which reduces labor and overhead costs.

The most effective is the first of these: reducing the number of synthetic steps. The reason for this is as follows. Cost is directly proportional to the overall yield; the yield, however, is exponentially and inversely proportional to the number of synthetic steps. For example, assuming an 80% yield for each step, a 10-step synthesis would give a total yield of 10%. By cutting the synthesis in half, to 5 steps,

Table 1.3. Specifications for Solvents and Heavy Metals

Solvent	Maximum, %	Metal	Maximum, ppm
Acetone	1	Iron	10
Benzene	ND[a]	Mercury	1
Chloroform	ND[a]	Palladium	4
Ethanol	2	Arsenic	2
Ethyl acetate	1	Boron	10
Heptane	1	Sodium	300
Methanol	0.05	Chromium	1
Dichloromethane	0.05	Lead	2
Toluene	0.1		

[a]Not detectable.

the overall yield would be 33%, or three times higher, and the cost three times lower, assuming similar costs for reagents.

Another strategy is to make the synthesis convergent. A 22-step linear synthesis at 80% yield per step would have an overall yield of 0.7%; if made convergent (i.e., two pieces made in 10 and 11 steps, respectively, are combined in step 22), the yield would be 10 times higher and the cost 10 times lower, if everything else is equal. Table 1.4 shows an example of how powerful and valuable process development can be; the chemistry of this example is described in Chapter 6.

In the "Introduction" I mentioned a 25-step research synthesis that started with D-glucose and cost $48,134 in chemicals to make 1 kg of drug substance. By some quick development (mostly reducing the amounts of solvents and reagents and improving the yields), the cost was brought down to $16,845/kg, or by a factor of 3 (see Table 1.4). Next we designed a new 17-step synthesis with L-malic acid as the starting material, which cost only $3,505/kg of drug substance, and finally an 11-step synthesis from L-mandelic acid, for only $1,439/kg. Whereas process optimization reduced costs by a factor of 3, we see that a new, shorter synthesis reduced the cost by another factor of 10.

To quote a *Harvard Business Review* article (7):

Although the pharmaceutical industry continues to be a business in which new product innovation is paramount, how a company manages process development will influence to a significant degree the extent to which it can dramatically lower its manufacturing costs and continue to excel at product innovation.

1.7. NEVER-ENDING STORY

Chemical development is generally considered a service operation to other departments; that means that we work under the restrictions of *deadlines* (somebody is always waiting for our processes or drug substance), *ideal process qualities* (which I have just discussed in great detail), and *targeted research* (molecules and problems are provided to us and must be solved, they cannot be revised or disregarded), while doing *research* that is unpredictable (it is impossible to predict a date by which a discovery will be made) and *development* that is open ended (there is always room for improving a process). We work with the best process available at the time, not with an ideal process.

Table 1.4. Lower Chemical Costs Through Process Development

Process	Development Stage	Relative Cost	Number of Steps	Starting Material
1	Preclinical	16,845	9 + 16 = 25	D-Glucose
2	IND	3,505	9 + 8 = 17	L-Malic Acid
3	Clinical	1,439	11	L-Mandelic Acid

1.8. EXAMPLES

Lest we forget that we are talking about chemistry, let me summarize this chapter with two real-life chemistry examples. The remainder of this book will further illustrate how one applies these principles.

Drug substance **1** (Fig. 1.1) was synthesized by the medicinal chemist as follows: in the last step of the synthesis, the triazine-oxide **2** was deoxygenated with palladium chloride; the triazine-oxide was prepared by a condensation of the oxime-hydrazone **3** with ethyl imidate **4**, at 100°C with about 40% yield; the hydrazone was prepared in two steps from 2-oxocineole (not shown). Let us concentrate on the last two steps. Keeping the criteria of an ideal process in mind, we notice three potential problems:

- Palladium chloride is a suspected carcinogen and is, therefore, a safety problem. Since it is used in the last step of a synthesis it is also a potential quality problem: Our specification for carcinogens in drug substances is not detectable, and this level may be difficult to achieve, to remove all of the palladium.
- The yield (40%) is low and is an economics concern.
- Most critical, the decomposition temperature of the hydrazone **3** was measured to be 110°C with an exotherm, and this temperature is too close to the reaction temperature (100°C) to be considered safe. This decomposition may be the reason for the low yield. The obvious solution, lowering the reaction temperature, did not work, as the reaction did not proceed at lower temperatures and the yield was even lower.

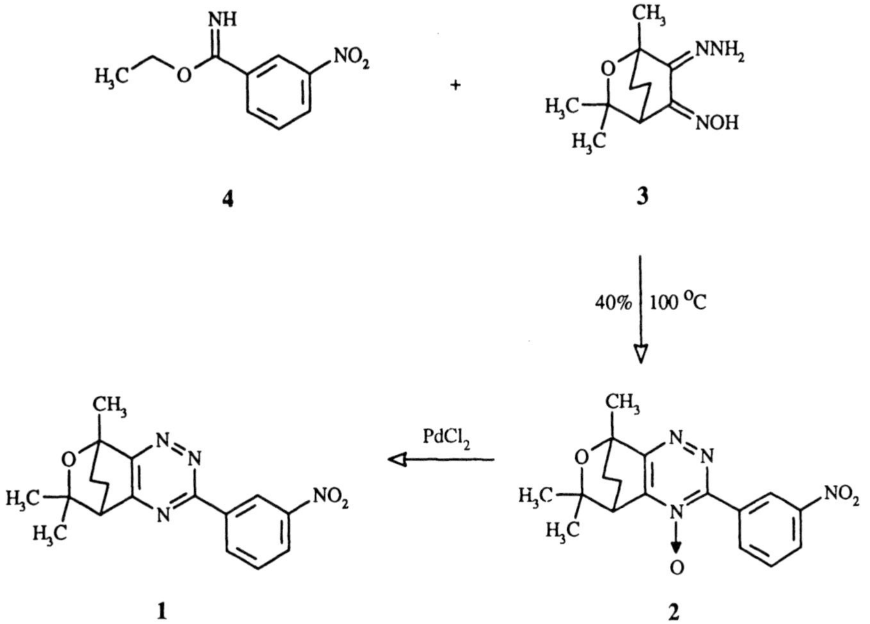

Fig. 1.1. Research synthesis of a drug substance.

When faced with so many problems in a short synthesis, the best strategy is to pursue a new synthesis. We did this by formally moving all the nitrogen-containing functional groups into the imidate (Fig. 1.2) (8). Ethyl imidate **4** was allowed to react with hydrazine to give *m*-nitrobenzimidic acid hydrazide **5**, which condensed with 2,3-dioxocineole **6** at room temperature to give drug substance **1** directly in 80% yield. In this way, all the mentioned problems have been avoided. Admittedly, hydrazine is also a carcinogen; however, it is not used in the last step, and it would be difficult to avoid hydrazine as a reagent in any synthesis of triazines (it was used in the previous synthesis as well, to make hydrazone **3**). The new synthesis is not shorter, alas, as it also requires two steps to make diketone **6** from 2-oxocineole.

The second example is our process development of the Simmons–Smith cyclopropanation **7** → **8** (Fig. 1.3) (9,10). Though it is the most useful method for making *cis*-cyclopropanes from olefins, this classic reaction cannot be easily scaled up for reasons of safety, reproducibility, and economics (all of which make it a non-ideal process). Specifically,

· Diethyl ether is considered an unsafe solvent.
· Zinc–copper amalgam is an expensive reagent.
· The reaction displays a delayed and violent exotherm (e.g., on a 22-L scale, all of the ether boiled off despite efficient condensers).
· The yields are low and not reproducible.
· The reaction foams excessively.

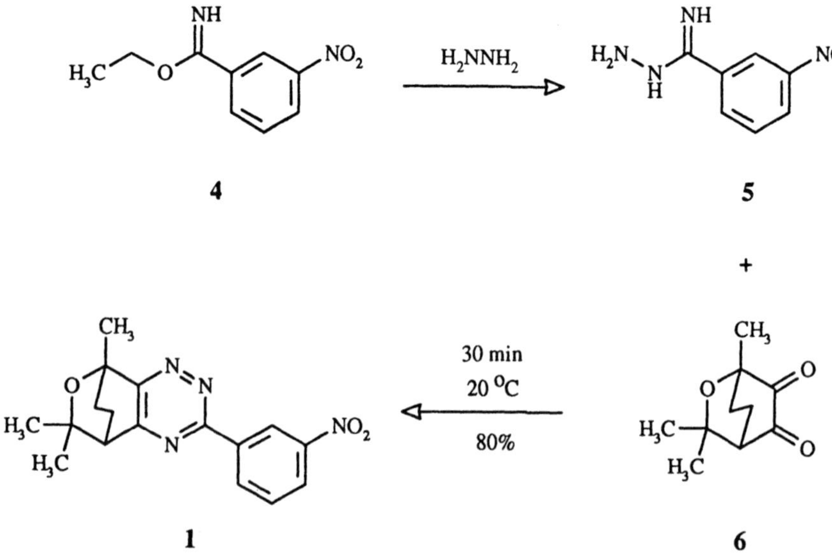

Fig. 1.2. New process for the drug substance.

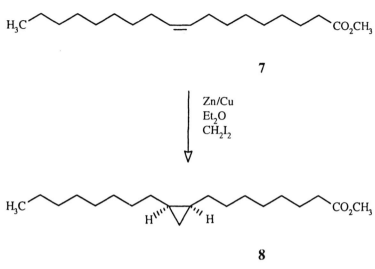

7

8

Fig. 1.3. Simmons-Smith cyclopropanation.

We solved all these problems by stepwise process modifications. The solvent was replaced with dimethoxyethane, which has the following advantages:

· Higher boiling point (85°C).
· Not as flammable (safety).
· Higher heat capacity (exotherm is more controllable).
· Higher reaction temperature (less delay in the start of reaction).

This also allowed the reagent to be replaced with zinc powder (cheaper). The addition of a catalytic amount of Red-Al (sodium bis(methoxyethoxy)dihydridoaluminate) avoids the delay in the start of the reaction, which presumably happens because the reagent

· Cleans the zinc surface of oxides.
· Removes any water in the solvent.
· Removes any peroxides in the solvent.

Any of these three can cause a slow start of the reaction.

The only problem left at this point was the nonreproducible yield (heterogeneous reaction). This was solved by conducting the reaction under ultrasonic irradiation, which resulted in extremely high and reproducible yields (around 90%); most losses were due to hydrolysis of the methyl ester in the workup. Finally, the zinc powder was replaced with two cones of zinc, suspended in the reaction mixture (11). This solid reduced the surface area of the zinc compared to the powder and resulted in a smoother and more controllable reaction (less foaming and the reaction can be controlled simply by pulling the zinc out of the reaction solution). This technology could be engineered into a plant design, if necessary.

BIBLIOGRAPHY

1. B. D. Naumann, E. V. Sargent, B. S. Starkman, W. J. Fraser, G. T. Becker, G. David Kirk, "Performance-Based Exposure Control Limits for Pharmaceutical Active Ingredients," *Am. Ind. Hyg. Assoc. J.,* **57,** 33–42 (1996).

2. N. I. Sax, ed., *Dangerous Properties of Industrial Materials,* 6th ed., Van Nostrand Reinhold, New York, 1984.

3. C. Reichardt, *Solvents and Solvent Effects in Organic Chemistry,* Verlag Chemie, Weinheim, Germany, 1988.

4. P. T. Anastas and C. A. Farris, eds., *Benign by Design. Alternative Synthetic Design for Pollution Prevention,* ACS Symposium Series **577,** American Chemical Society, Washington, D.C., 1994.

5. P. T. Anastas and T. C. Williamson, eds., *Green Chemistry: Designing Chemistry for the Environment,* ACS Symposium Series **626,** American Chemical Society, Washington, D.C., 1996.

6. ICH, *Residual Solvents,* International Committee on Harmonization Guideline, Geneva, Switzerland, 1997.

7. G. P. Pisano and S. C. Wheelwright, "The New Logic of High Tech R & D," *Harvard Bus. Rev.,* Sept.–Oct., 1995, p. 101.

8. O. Repič, P. G. Mattner, and M. J. Shapiro, "Preparation of New 1,2,4-Triazines." *J. Heterocycl. Chem.,* **19,** 1201–1204 (1982).

9. O. Repič and S. Vogt, "Ultrasound in Organic Synthesis: Cyclopropanation of Olefins with Zinc-Diiodomethane," *Tetrahedron Lett.,* **23,** 2729–2732 (1982).

10. U.S. Pat. 4,472,313 (Sept. 18, 1984), "Cyclopropanation of Olefins," U. Giger and O. Repič (to Sandoz).

11. O. Repič, P. G. Lee, and U. Giger, "Large-Scale Cyclopropanation," *Organ. Prep. Proced. Int.,* **16,** 25–30 (1984).

2

IMPURE THOUGHTS

In Chapter 1 we noted that it was the job of the process chemist to ensure that the drug substance meets quality specifications. If impurities exceed the limits, the process chemist must minimize these by-products or else add purification steps.

A most exciting activity in chemical development is troubleshooting impurities, because this task is intellectually challenging, like playing detective. An unknown impurity appears in the final product that causes it to fail specifications; what now? The strategy is to:

- Isolate the by-product by prep high-pressure liquid chromatography (HPLC).
- Determine its structure by NMR and mass spectroscopy (MS).
- Propose a mechanism for its formation.
- Identify the synthetic step in which it forms.
- Alter the reaction conditions, using the mechanistic hypothesis, to minimize the formation of the undesirable by-product.

Twenty examples of this general procedure follow.

2.1. EXAMPLE 1

A major objective was to eliminate chromatography from the synthesis of **1** (Fig. 2.1) by developing an alternative process to produce this drug substance in high purity (1,2). The chromatography was used mainly to remove triphenylphosphine oxide, a by-product of the Wittig reaction, and a dimeric by-product.

The solution was to switch to the Horner–Emmons reaction conditions for the coupling of the two halves of the molecule (Fig. 2.2). This change avoided the tri-

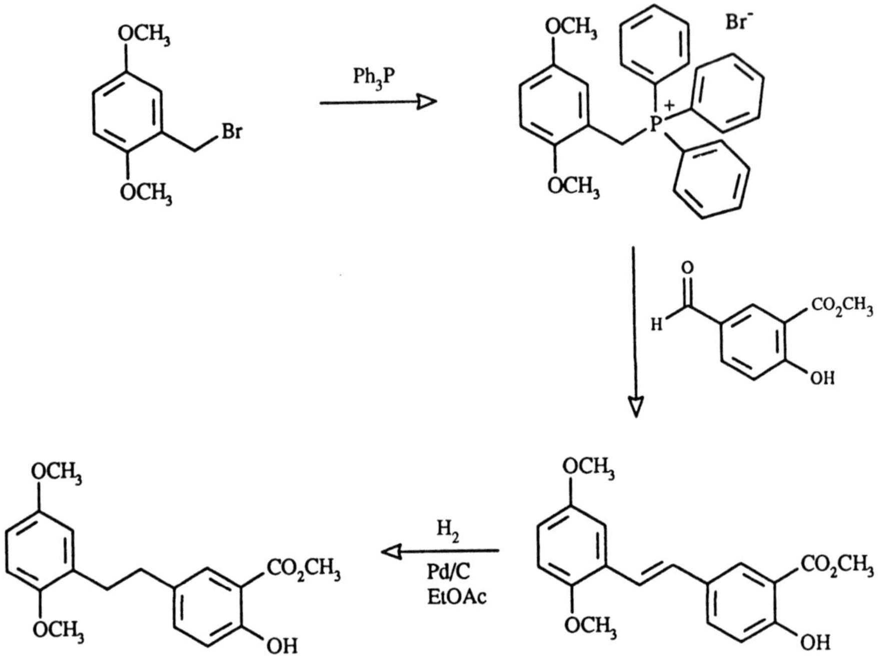

Fig. 2.1. Synthesis via the Wittig reaction.

phenylphosphine oxide by-product; on the other hand, the dimeric impurity **2** and several other by-products (**3–5**) were observed in the drug substance (Fig. 2.3). They were tracked down and eliminated as follows.

2.1.1. By-Products 2 and 3

The same dimeric by-product **2** had already been noticed in the original synthesis and can be explained as follows. In the bromination step, the stabilized benzylic cation undergoes either an aromatic electrophilic substitution or an *O*-alkylation of the 2,5-dimethoxybenzyl alcohol to form by-products **6** and **3**, respectively (Fig. 2.4). Both are carried through the synthesis: ether **3** remains unchanged, whereas the dimer **6** participates in all the reactions, ending as by-product **2** in the drug substance (Fig. 2.5).

The formation of ether **3** (see Fig. 2.4) could be completely suppressed and the amount of dimer **2** halved by reversing the addition of the reactants (the alcohol was added to phosphorus tribromide so that there was never an excess of the alcohol present in the reaction mixture) and by simultaneously lowering the reaction temperature to −5°C. Precipitation of the crude product from ethyl acetate–heptane produced pure 2,5-dimethoxybenzyl bromide.

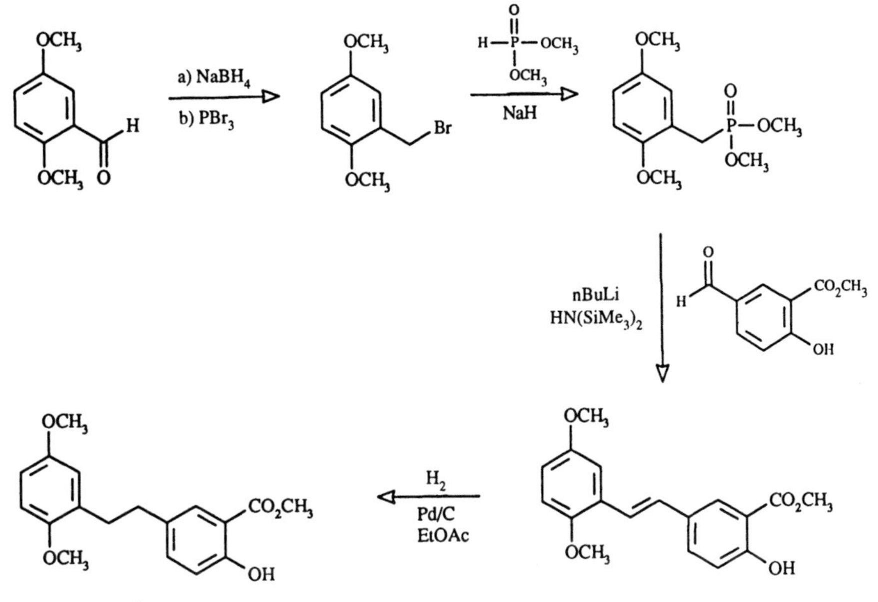

1

Fig. 2.2. Synthesis via the Horner–Emmons reaction.

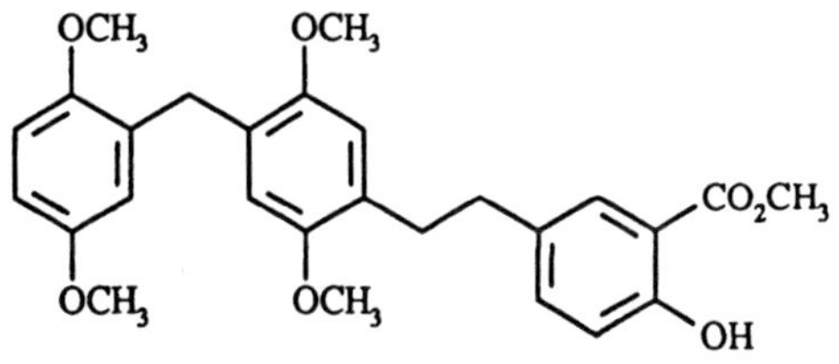

2

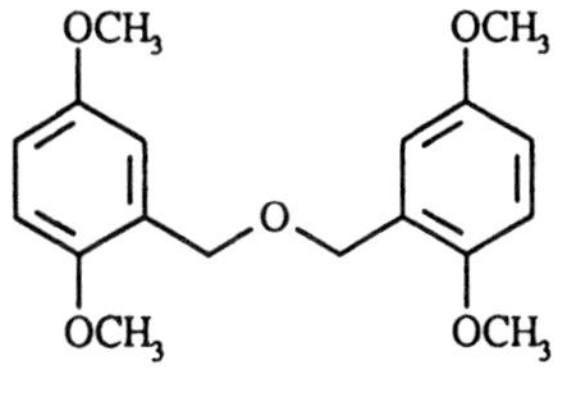

3

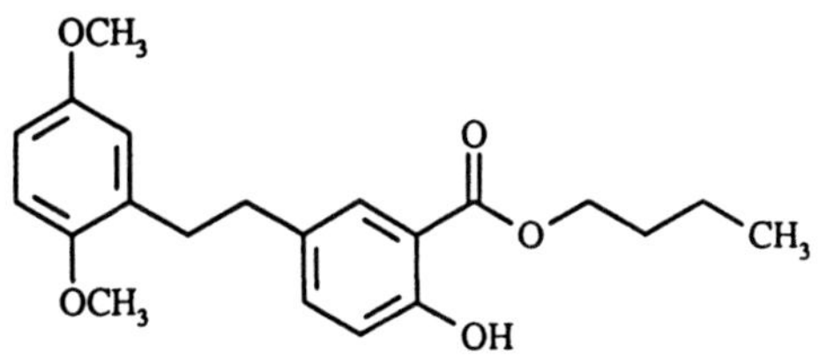

4

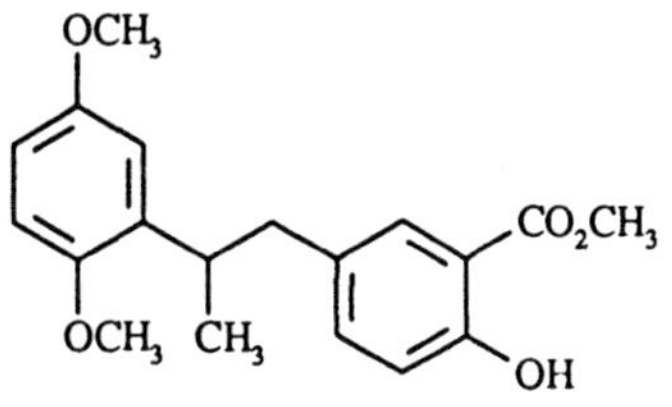

5

Fig. 2.3. Structures of the by-products.

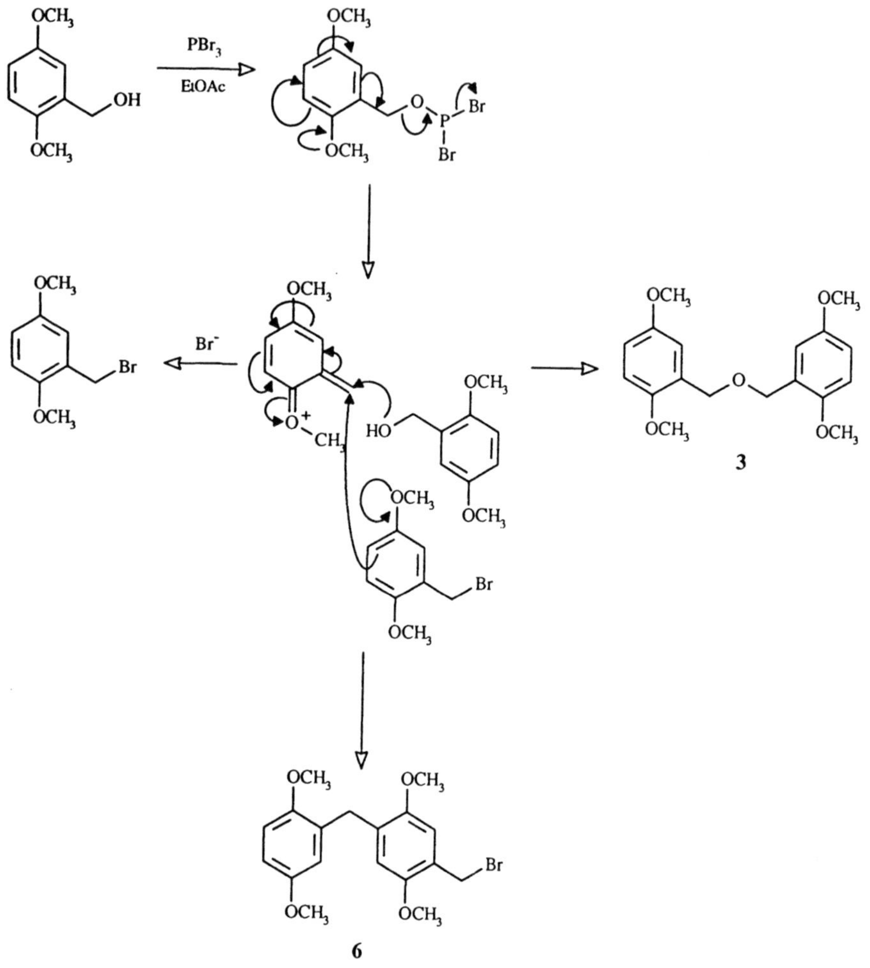

Fig. 2.4. Mechanism for the formation of two by-products.

2.1.2. By-Product 4

The *n*-butyl ester (by-product **4** in Fig. 2.6) was easily traced to the Horner–Emmons reaction, in which the methyl ester transesterified with lithium butoxide, an impurity in *n*-butyllithium generated by air oxidation of the reagent. The amount of this impurity varied, depending on the batch of *n*-butyllithium used, but it could be reduced by using fresh reagent and by recrystallization of the drug substance.

2.1.3. By-Product 5

Last, another trace impurity (ca. 0.7%) was identified in the recrystallized drug substance as the methylated by-product **5** (Fig. 2.7). The methylating agent is probably

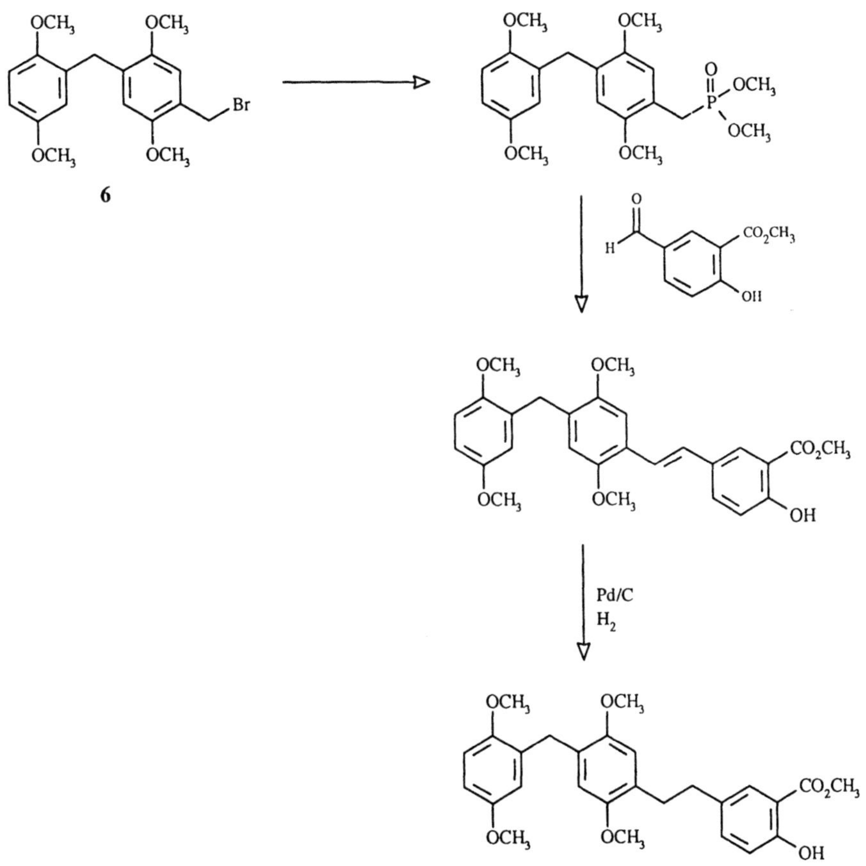

Fig. 2.5. By-product is carried through the synthesis.

the dimethyl phosphonate in the Horner–Emmons reaction, where it methylates the phosphonate α-anion, and the methylated product is carried through all the subsequent reactions.

The methylated impurity was minimized by changing the order of addition of the reactants: The aldehyde and phosphonate were added *simultaneously* to the base, limiting the exposure of the phosphonate anion to any competing reactions (methylation, oxidation, protonation). The reaction temperature was also lowered to −45°C, again to slow any side reactions of the phosphonate anion.

The new process produced the drug substance—after two crystallizations and with no chromatography—in a 67% yield; it contained 0.3% of the methylated by-product **5** as the only impurity.

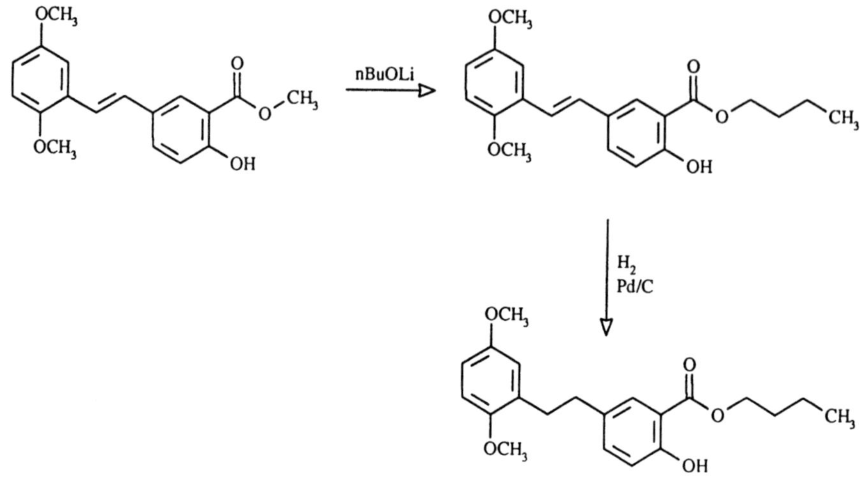

Fig. 2.6. Transesterification with *n*-butoxide.

2.2. EXAMPLE 2

As in the previous example, HPLC analysis of drug substance **7** (Fig. 2.8) revealed several by-products of the synthesis (3). They were isolated and identified as **8, 9,** and **10** (Fig. 2.9). Interestingly, compound **10** was a single diastereomer; this fact has been confirmed by synthesis of both isomers starting with (*S*)-*α*-methylbenzylamine and (*R*)-*α*-methylbenzylamine, respectively (4). The mechanism of its formation is speculative, but it is probably a base-catalyzed intramolecular diastereoface-selective methyl migration (Fig. 2.10); the selectivity arises because one face is blocked by the bulky naphthyl group. Even though it is less likely (because of the apparent high diastereoselectivity), the mechanism could also be an intermolecular demethylation and methylation via CH_3Cl. Optimizing the reaction conditions reduced the amount of the by-product **10** from 2 to 0.1%. The changes included prebasifying HCl salt **11** (the coupling step then requires less base and contains less chloride, either of which could cause the base-catalyzed methyl migration), and using less *i*-butyl chloroformate (which can acylate amide **12** and increase the acidity of the benzylic proton).

Impurity **8** was easier to explain: The raw material benzyl-*N*-methylamine was contaminated with benzylamine. Because of time constraints, we were not able to find a source of pure benzyl-*N*-methylamine. Instead, we treated it with 3% of ethyl trifluoroacetate at 0°C before use, which reacted preferentially with the primary amine, thus blocking its further participation in the synthesis (5).

The *i*-butyl carbamate impurity **9** was undoubtedly formed by acid-catalyzed loss of the *t*-butyl carbamate and subsequent reaction with *i*-butyl chloroformate (Fig. 2.11). The impurity was minimized by three synergistic solutions: (*1*) by using an excess of 4-methylmorpholine base, (2) by precipitating intermediate **13** so that

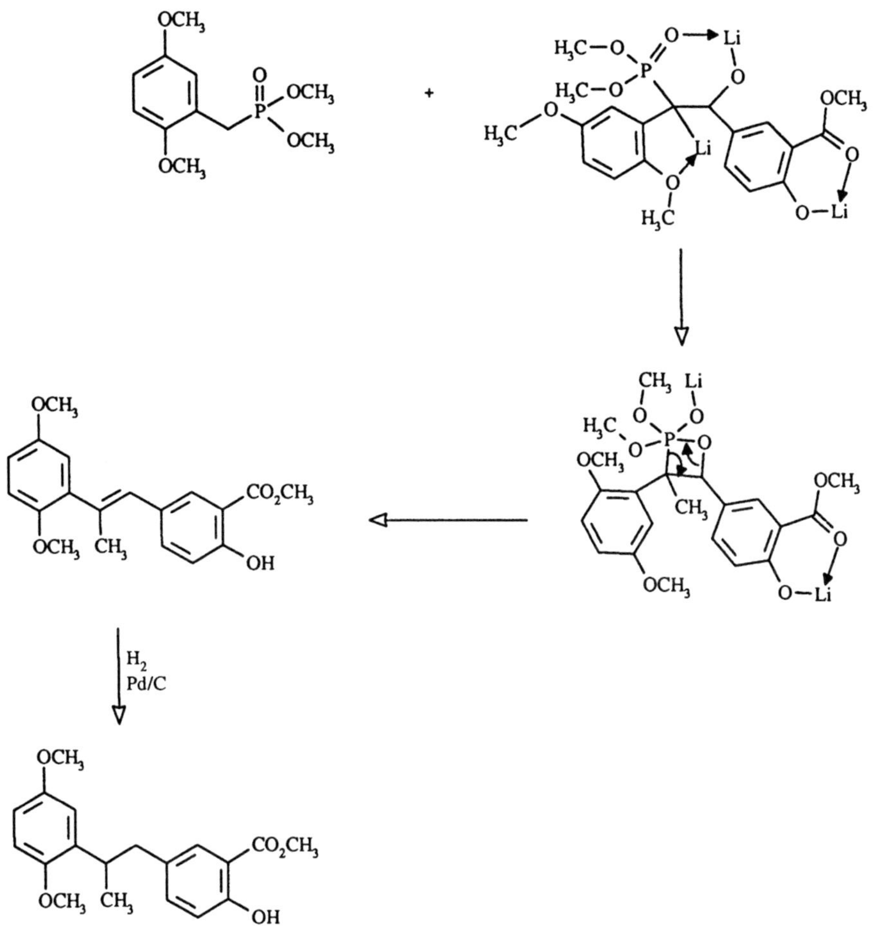

Fig. 2.7. Dimethyl phosphonate as a methylating agent.

the impurity ended in the filtrate, and (*3*) by prebasifying the HCl salt of **11** with aqueous sodium hydroxide. The latter operation removed all acid catalysts from the reaction but was performed also for other reasons: It was easier to add **11** to a reactor as a liquid free-base than as a solid salt, and the basification with NaOH removed some β-naphthyl-L-alanine, another impurity. Simultaneously, two unacceptable solvents in the synthesis—dichloromethane and dioxane—were replaced with ethyl acetate.

2.3. EXAMPLE 3

The unusual drug substance **14** (Fig. 2.12) was prepared in 30% yield by alkylating the octadecyl sulfide **15** in dichloromethane with *n*-propyl iodide in the presence of silver

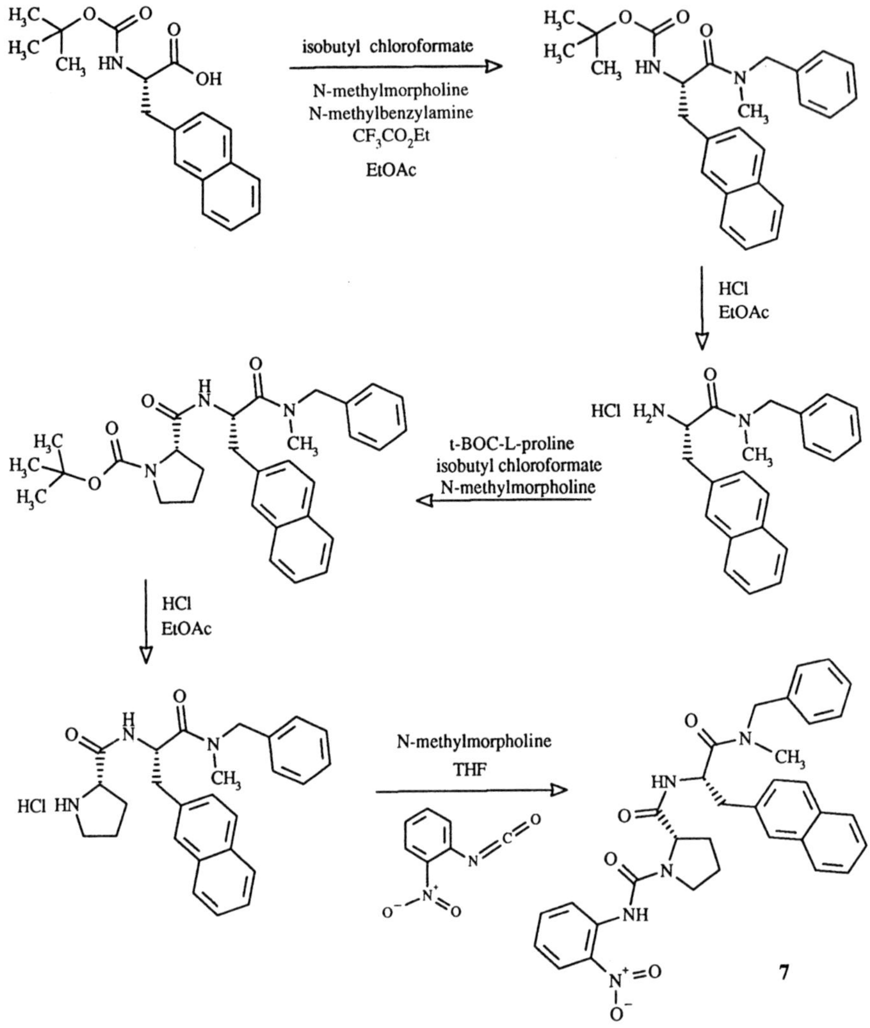

Fig. 2.8. Synthesis of SDZ NKT 343.

tetrafluoroborate (6). Improving the yield required extensive experimentation. The solvent was varied—Me$_2$SO, hexamethylphosphoric triamide (HMPT), N,N–dimethylformamide (DMF), dimethoxyethane (DME), nitromethane, diglyme, dichloroethane, dichloromethane)—but no improvement was observed, and dichloromethane was kept. Using an *excess* of n-propyl iodide and silver tetrafluoroborate did not help either. In a blank experiment, the reagents seemed to react among themselves, probably producing AgI, HBF$_4$, and propene. The reaction mixture was indeed quite acidic (HBF$_4$?), although no gas (propene) evolution was observed. Propene probably formed a complex with AgI.

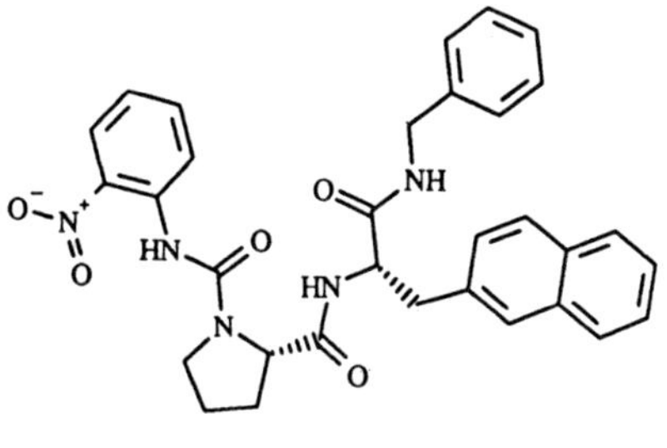

8

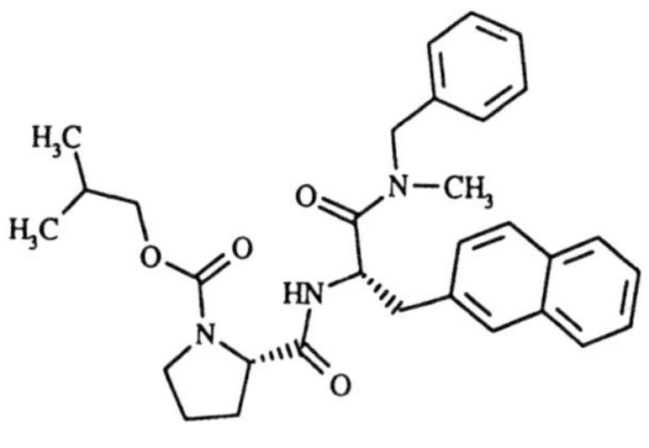

9

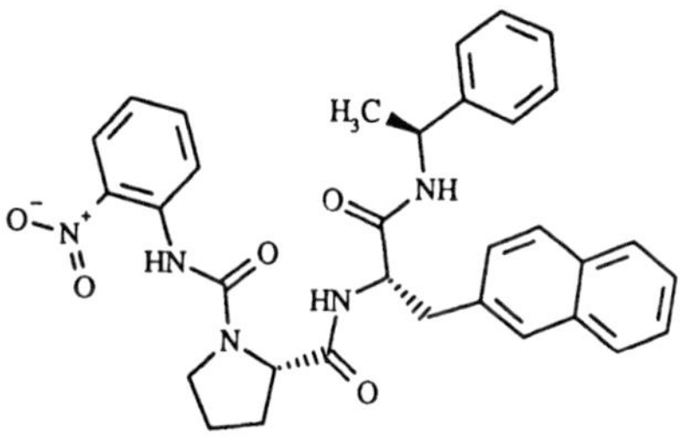

10

Fig. 2.9. Chemical structures of three by-products.

Even more perplexing than the low yield was the observation by ^{13}C NMR that up to 50% of the product was not the *n*-propyl but the isopropyl sulfonium isomer. The reagent *n*-propyl iodide exhibited a clean NMR spectrum, free of isopropyl iodide, so the raw material was not the source of the impurity (the reagent was, however, deeply colored, indicating the formation of iodine). Furthermore, the *n*-propyl group in the drug substance could not be persuaded to isomerize under any conditions tried: recrystallized from warm (45°C) ethanol three times, chromatographed through silica gel twice, melted on a steam bath, resubjected to reaction conditions for 20 h, or stirred with 1 equiv of iodine for 2 h. So iodine was not the culprit; the isomerization of *n*-propyl iodide must have occurred during the reaction.

Here, the solution to the problem came serendipitously and not by predicting the mechanism: While trying to optimize the yield, toluene was also tried as a solvent; not only did the yield jump from 30 to 60% but also no isopropyl isomer was formed. The yield was further improved to 85% by using 3 equiv of *n*-propyl iodide and 1.1 equiv of silver tetrafluoroborate. The factors noticed with dichloromethane (excess reagent or impure reagent) did not produce any isomer if toluene was used as the reaction solvent. The mechanism of this solvent effect is not proven; perhaps toluene complexes with silver and attenuates its Lewis acidity, which prevents the formation of the secondary propyl cation (*H*-shift to form the more stable carbonium ion). At the same time, toluene probably diminishes the complexation of Ag$^+$

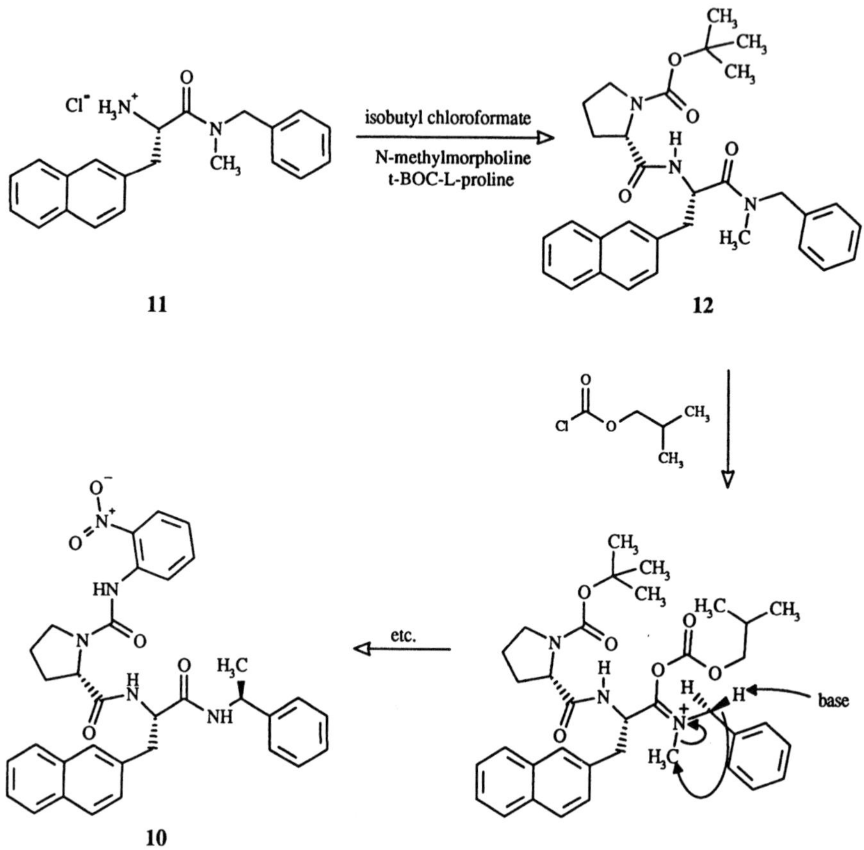

Fig. 2.10. A mechanism for methyl migration.

with the sulfide and speeds up the desired reaction. This example of a solvent effect illustrates why a solvent change is a significant process change (see Chapter 8).

2.4. EXAMPLE 4

Speaking of isopropyl groups, we encountered another mystery with an isopropyl group, but the explanation in this case was much simpler. In the early phases of developing the synthesis of Lescol (see Fig. 3.30, Chapter 3), we noticed, in one batch, a few percent of an impurity, which by NMR turned out to be the N-ethylindole homolog, instead of the desired N-isopropylindole (7,8). After excluding all possible mechanisms of a demethylation or transalkylation during the synthesis, we traced the impurity to the starting material, N-isopropylaniline. Although the problem and the solution became trivial (we found a source of pure N-isopropylaniline, free of other anilines), the issue of unexpected impurities in the drug substance is

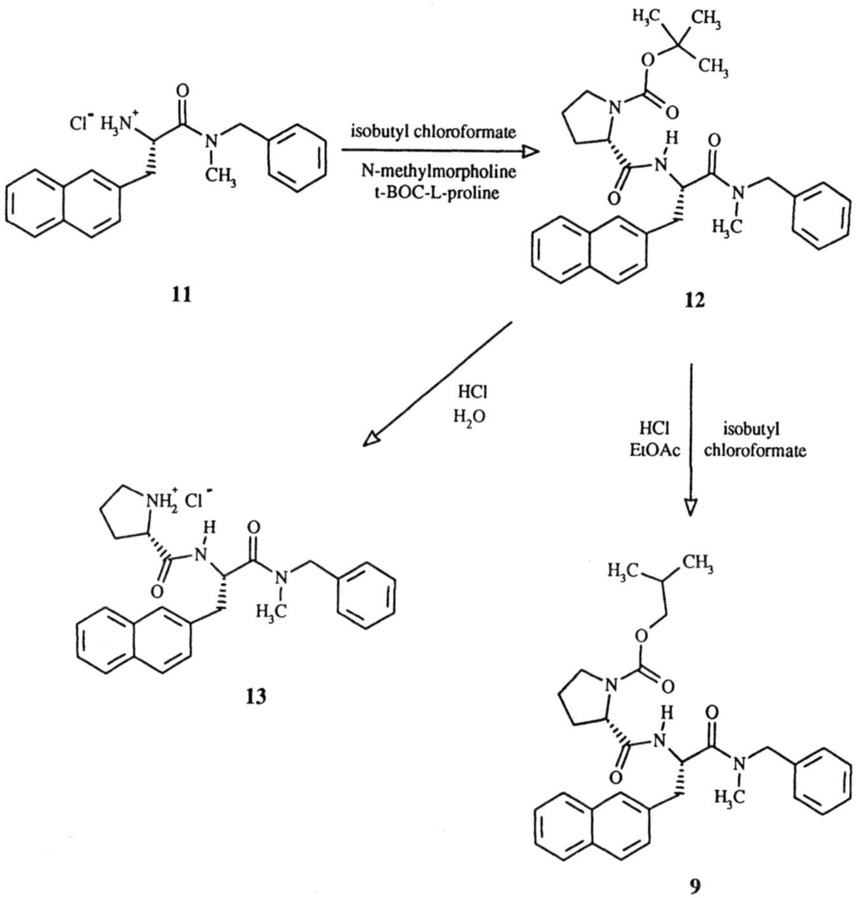

Fig. 2.11. Formation of by-product **9**.

always confounding at first and yet exhilarating when a solution is found. Nevertheless, this example points out that a change of the raw material supplier can be a significant process change (see Chapter 8), because it can affect the quality of the drug
substance, in spite of the many subsequent steps and purifications. This is especially true for enantiomers, homologs, and isomers, which resemble the drug substance and can easily be carried through the synthesis undetected.

2.5. EXAMPLE 5

Similar to Example 1, in an early convergent synthesis of fluvastatin (see Fig. 3.16,
Chapter 3), a large amount of a methylated by-product **16** was observed (Fig. 2.13)
(9). This by-product could not be removed in subsequent steps; however, considera-
tion of the mechanism of its formation (methyl phosphonate **17** in the Horner–

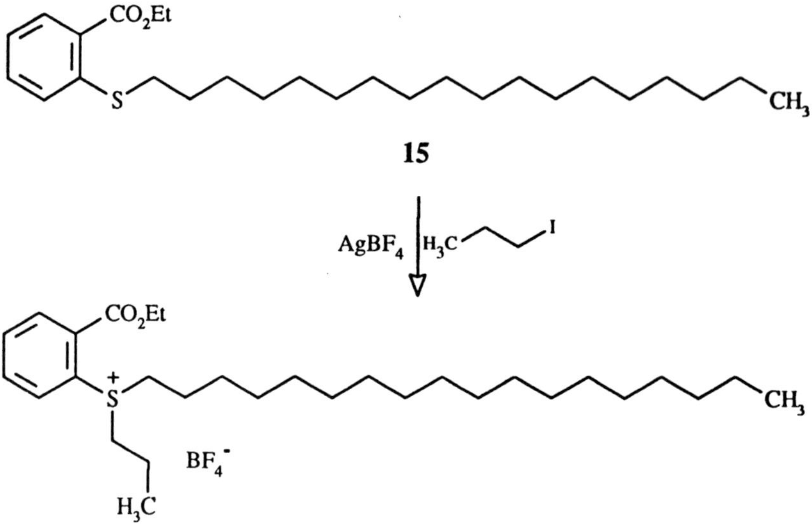

15

AgBF$_4$ H$_3$C⟍⟋⟍I

BF$_4^-$

H$_3$C

14

Fig. 2.12. Synthesis of SDZ 56-472.

Emmons step is a methylating agent) quickly led to the solution of the problem (10): by adding the phosphonate to *n*-butyllithium (reverse addition), no excess phosphonate is present and thus it cannot methylate the phosphonate anion (Fig. 2.14). As we have now seen several times, the order of addition of reactants must be considered a significant process variable, as it can affect the purity of the drug substance (see Chapter 8).

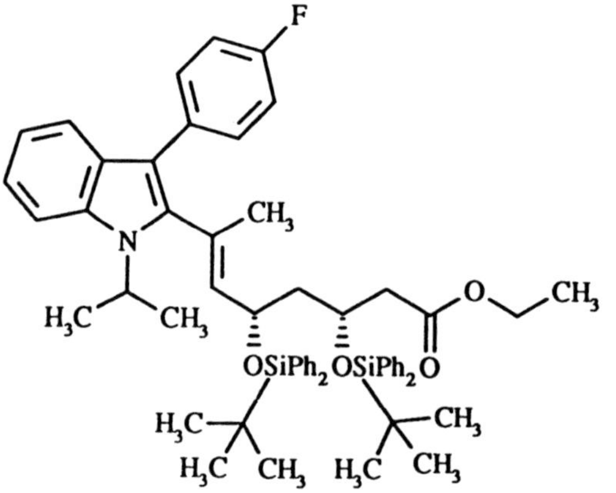

16

Fig. 2.13. Structure of methylated by-product.

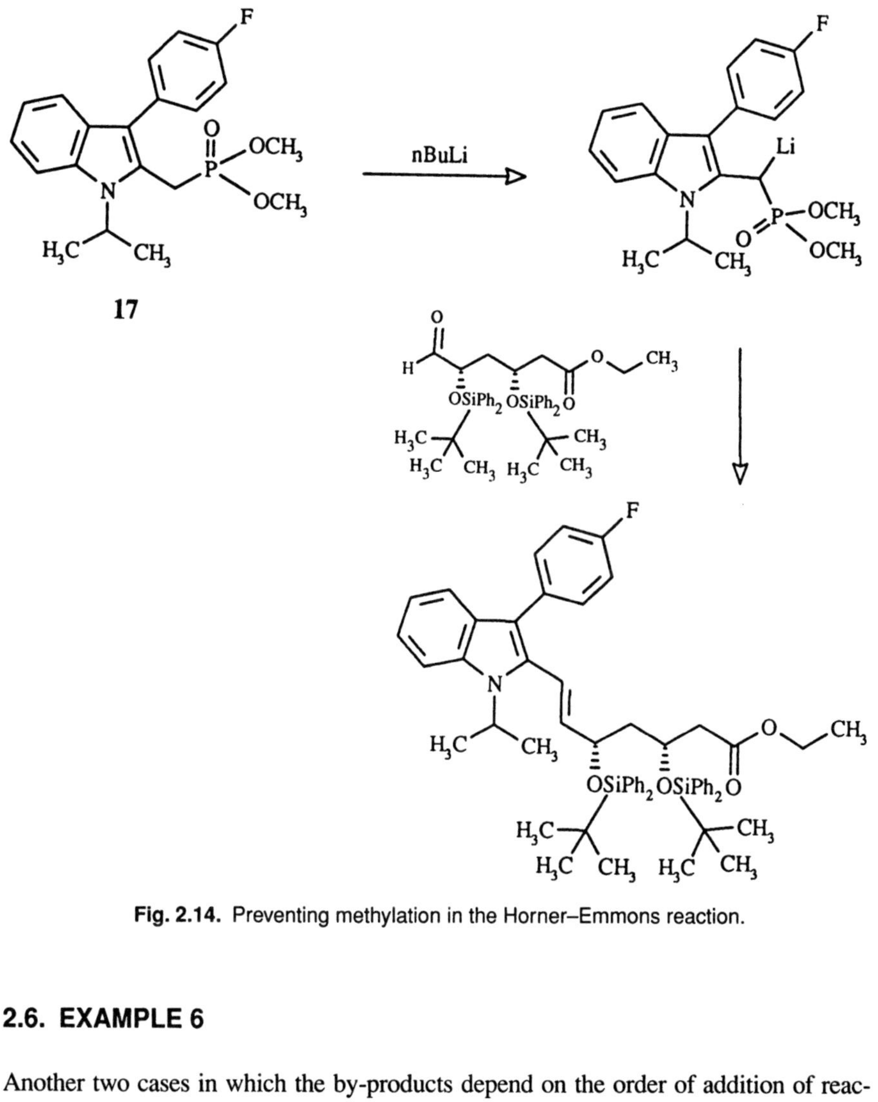

Fig. 2.14. Preventing methylation in the Horner–Emmons reaction.

2.6. EXAMPLE 6

Another two cases in which the by-products depend on the order of addition of reactants, will be described here. An indene analog of fluvastatin (the synthesis of the latter is shown in Fig. 3.30), was synthesized similarly (11). As in the case of fluvastatin, the required conjugated aldehyde intermediate **21** was made in only three steps (Fig. 2.15) (12). As simple as these three steps seem, they required considerable development time, because of the formation of side products, poor yields, and isolation difficulties.

In Step 1, the addition of *p*-fluorophenyl Grignard reagent to indanone **18** gave dimer **22** as the major product: The Grignard reagent acted as a base and caused enolization and self-condensation of the indanone (Fig. 2.16). This by-product could be avoided by reversing the order of addition of reactants and by conducting the addition slowly at room temperature or lower. Sulfuric acid was then needed to dehydrate the intermediate addition product to give **19**.

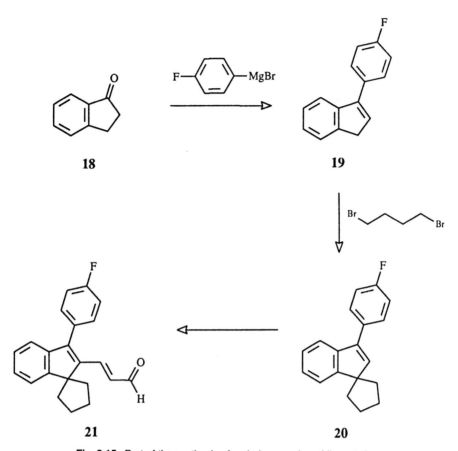

Fig. 2.15. Part of the synthesis of an indene analog of fluvastatin.

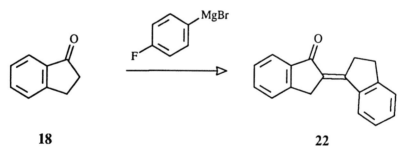

Fig. 2.16. Formation of one by-product.

Step 2, the alkylation of indene **19** with 1,4-dibromobutane (see Fig. 2.15), also caused problems because of a side-reaction, namely alkylation in the undesired position, leading to **23** instead of **20** (Fig. 2.17). The side reaction **19** → **23** was minimized, but not eliminated, by optimizing the base, solvent, and reaction temperature. The best conditions were NaH, tetrahydrofuran (THF) at reflux, and inverse addition; the yield was thus increased to 68%.

The vinyl formylation **20** → **21** proceeded in high yield, but the product had to be crystallized twice, because of impurities from the previous step, for a final yield of 70%.

2.7. EXAMPLE 7

Another analog of fluvastatin, a pyrimidine, was synthesized similarly, but this time the problem was in the vinyl-formylation step **26** → **27** (Fig. 2.18) (13,14). Thus *p*-fluoroacetophenone **24** was condensed with methyl isobutyrate, and the resulting 1,3-diketone **25** was allowed to react with dimethylguanidine to give the pyrimidine **26** in only two steps. The conjugated aldehyde side chain was attached in one step, using *N*-methyl-*N*-phenylaminoacrolein, as in the fluvastatin process (see Fig. 3.30). The yield in this case was low, however; and about 50% of the starting material was always recovered. Unreacted starting material can be considered an impurity for all practical purposes, which is why this example is mentioned here. A 50% recovery of starting material is often a clue for inhibition of a reaction by the product. For example, a reaction by-product (HCl?) may be blocking the starting material **26** from further reaction by protonating the dimethylamine, deactivating the pyrimidine, and switching off the desired electrophilic aromatic substitution reaction.

Using this hypothesis, a variety of bases was evaluated: triethylamine, potassium carbonate, Proton-Sponge, potassium *t*-butoxide, basic alumina, diisopropylethylamine, tetramethylguanidine, 1,8–diazabicyclo[5.4.0]undec-7-ene (DBU), 4-dimethylaminopyridine (DMAP), Amberlyst A-21, sodium carbonate, lithium carbonate, imidazole, 4-pyrrolidinopyridine, propylene oxide, tetramethylethylenediamine, and so forth. The solution was not simple. The base had

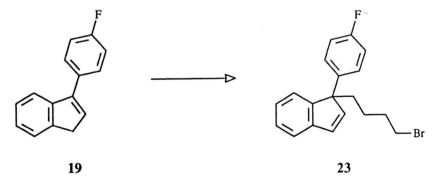

19   **23**

Fig. 2.17. Formation of the second by-product.

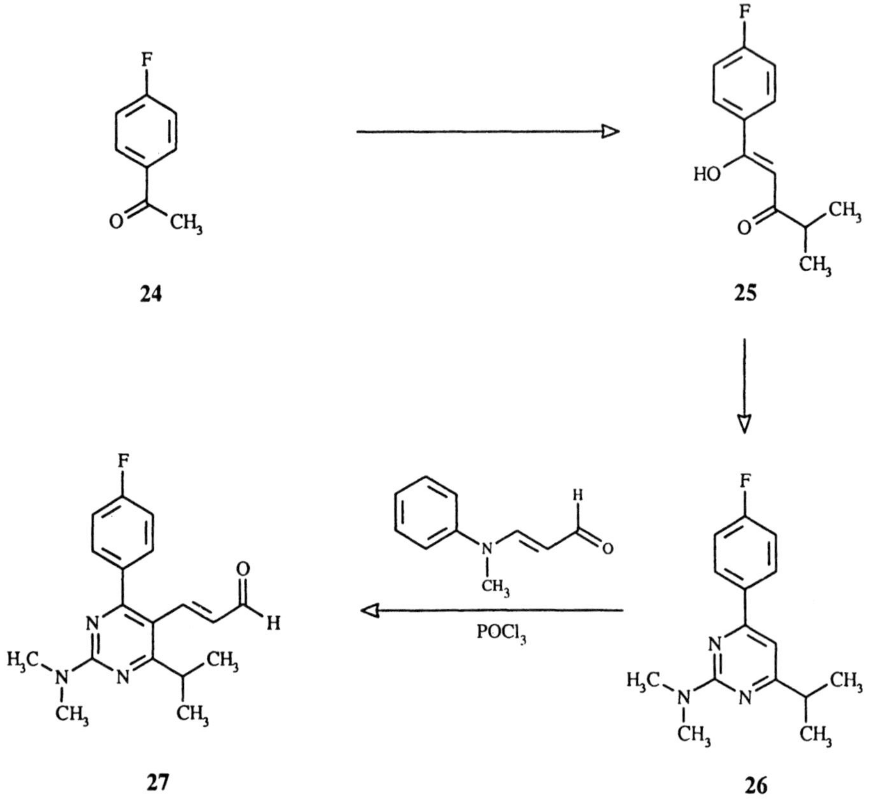

Fig. 2.18. Part of the synthesis of a pyrimidine analog of fluvastatin.

to be stronger than the reactant, aminopyrimidine; however, if too strong or too nucleophilic, the base would react with the activated iminium reagent. Using an extra equivalent of the starting material to act as the base was successful, but did not, of course, improve the yield. The breakthrough and the solution were found to be in the use of 2,3,5,6-tetramethylpyrazine, which apparently has just the right pK_b, as the yield of **27** jumped to 78%.

2.8. EXAMPLE 8

In the synthesis of another 3-hydroxy-3-methylglutaric acid (HMG) Coenzyme A reductase inhibitor, this one a naphthalene analog, the *ortho*-directing oxazoline group in **28**, having served its purpose, required conversion to aldehyde **30** (Fig. 2.19) (15). Initially, this was done by reducing **28** to oxazolidine **29**, which could be hydrolyzed to the aldehyde **30**. Both of these steps, however, gave a low yield, because of the formation of a considerable amount of by-products. The structure of the respective by-products and the possible mechanism of their formation pointed to the solution of the problem in each case.

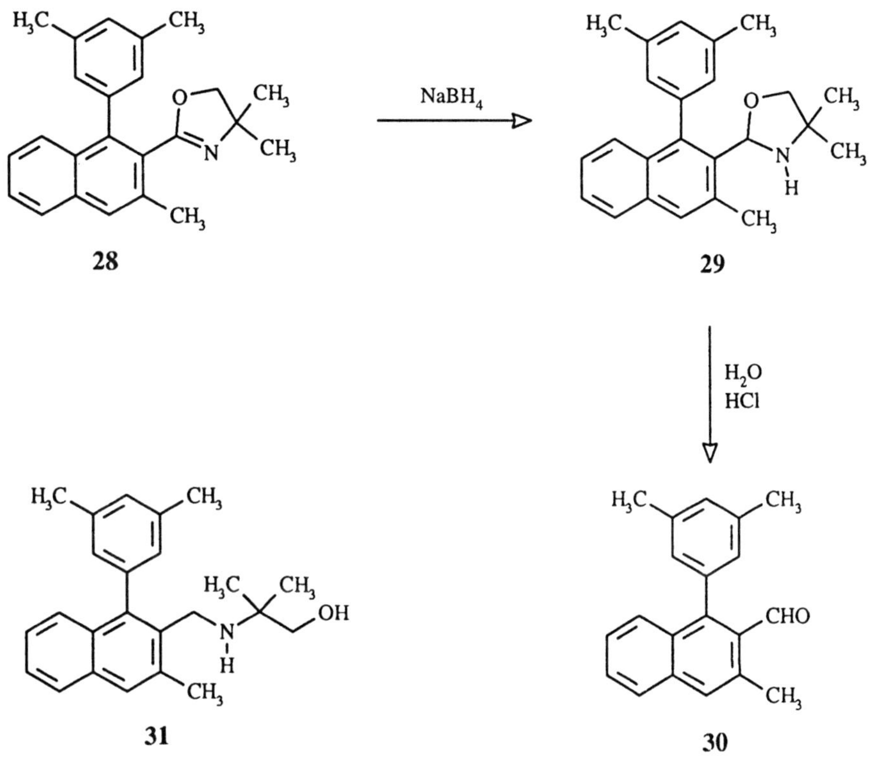

Fig. 2.19. Part of the synthesis of a naphthalene analog of fluvastatin.

The by-product of reaction **28** → **29** was **31**, a result of overreduction. Thus employing less hydride in a less-polar solvent than *N,N*-dimethylacetamide minimized the overreduction.

The by-product of reaction **29** → **30** was more intriguing: It had the same molecular ion (by mass spectroscopy) as the desired product **30** but lacked an aldehyde proton and one aromatic proton (by NMR). The structure determined for this by-product was **32**, and a possible mechanism for its formation is an intramolecular electrophilic aromatic substitution by the imine **33**, which is an intermediate leading to the desired aldehyde **30** (Fig. 2.20). The resulting amine **34**, being doubly benzylic, is then solvolyzed in the acidic aqueous medium of the reaction to give **32**. By employing milder conditions for the hydrolysis (silica gel), the formation of side product **32** was entirely avoided.

2.9. EXAMPLE 9

Drug substance **35** was synthesized from three precursors **36–38** (Fig. 2.21) (16). Its analysis showed the presence of bromine, however. The problem was traced to

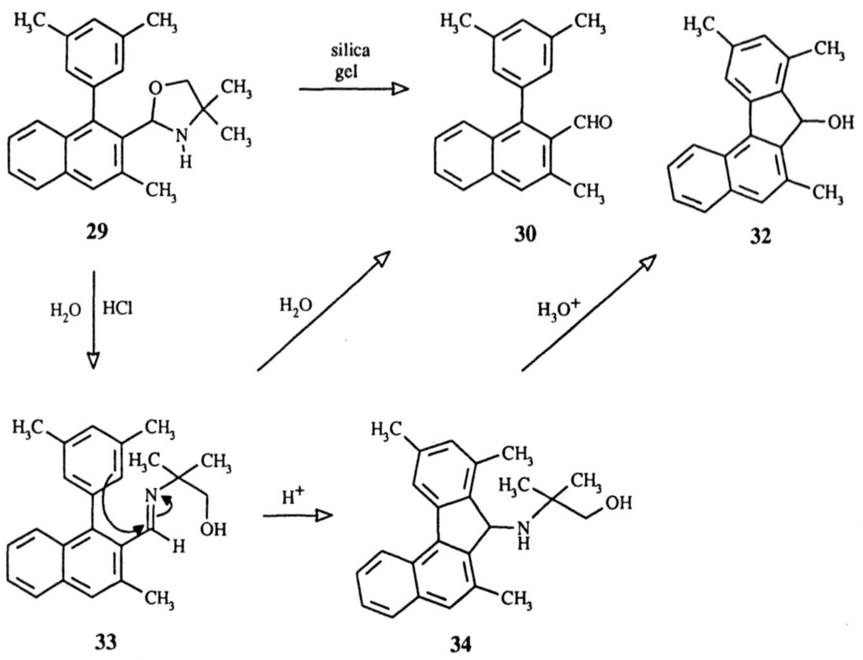

Fig. 2.20. Mechanism for by-product formation.

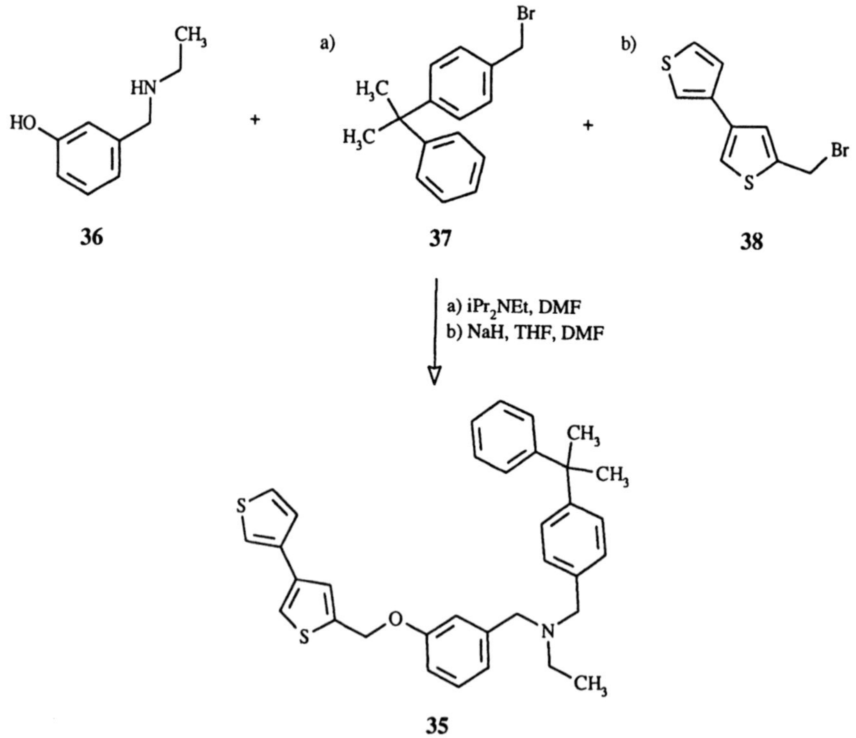

a) iPr₂NEt, DMF
b) NaH, THF, DMF

Fig. 2.21. Synthesis of SDZ 281-915.

the preparation of **37** (Fig. 2.22): The bromination of 2-phenyl-2-*p*-tolylpropane with *N*-bromosuccinimide (NBS) gave dibromide **38** as a by-product, which in turn gave immonium bromide **39** after substitution of the other bromine by amine **36**.

Having identified the cause of the problem, the solution was easily devised (17). This time, the solution was a purification: A high-vacuum fractional distillation separated **37** from **38** and gave the monobromide **37** in >98% purity, and—two steps later—the drug substance **35** was free of bromine.

2.10. EXAMPLE 10

Described here is another example of a halogenated impurity. Compound **41** (Fig. 2.23) had been identified as an oxidation by-product of dihydroergotamine and was required as an analytical reference (18,19).

The synthesis was achieved in six steps, employing calcium hypochlorite in the key step for oxidizing 6-methylergoline **42** to the oxyindole **43** (Fig. 2.24) (20). Several other oxidation conditions (NBS, Me$_2$SO, potassium persulfate, manganese dioxide) led to little or no product. A major by-product of the hypochlorite oxidation, however, was detected by MS and identified as the chlorinated product **44**. To eliminate this by-product, a hydrogenation step was added (palladium on carbon) to reduce the overoxidized (chlorinated) by-product **44** to the desired **43**.

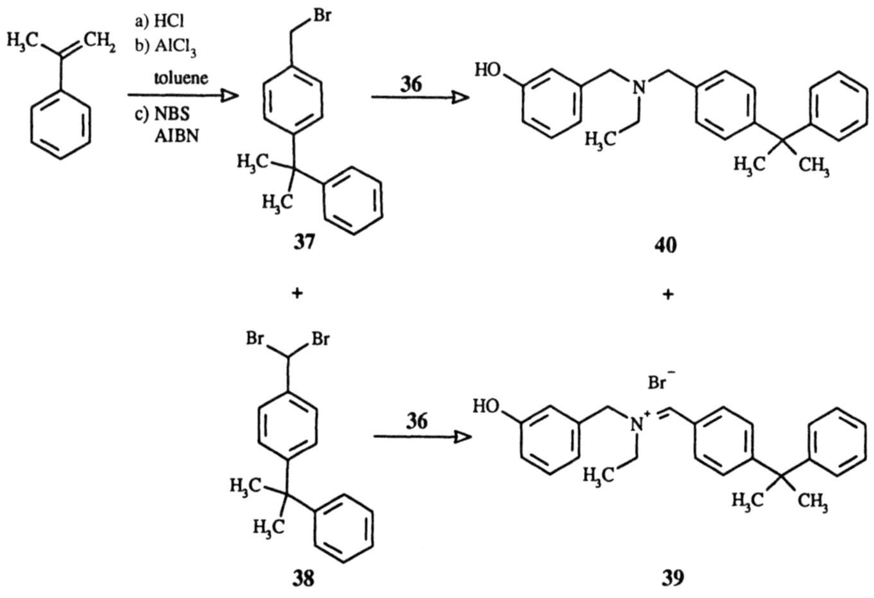

Fig. 2.22. Formation of brominated by-product.

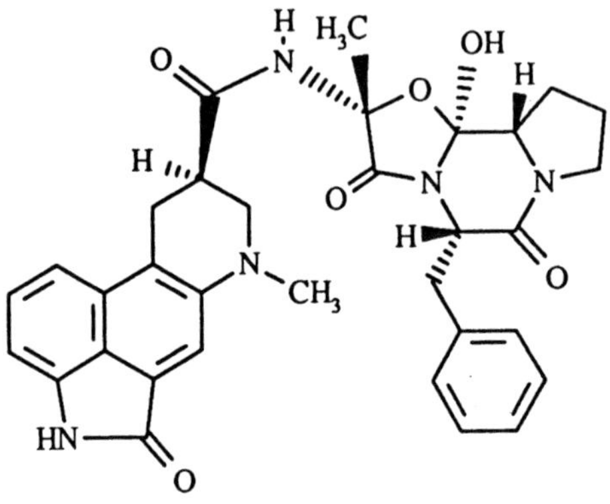

41

Fig. 2.23. Oxidation by-product of dihydroergotamine.

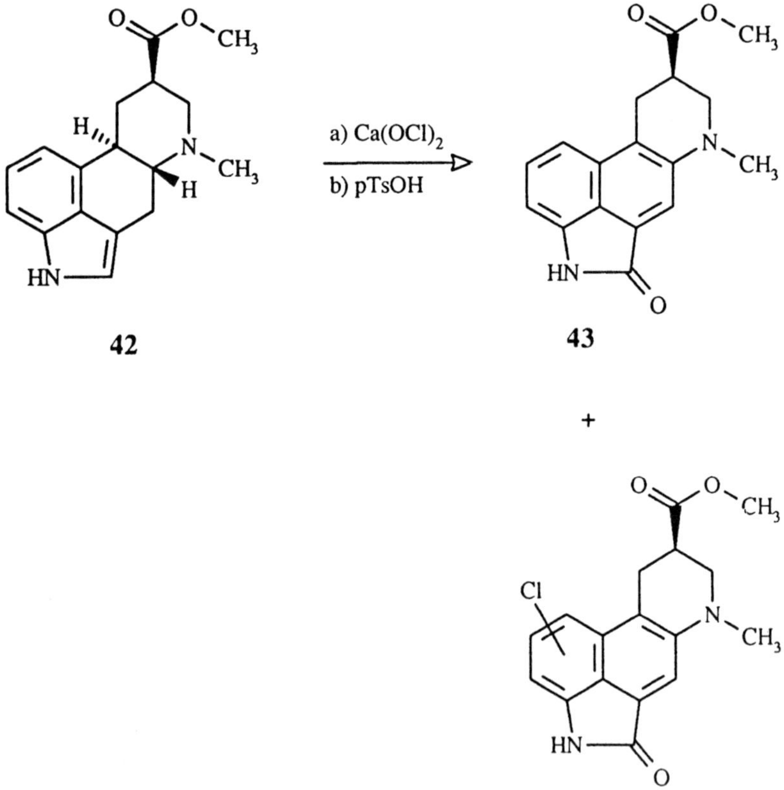

42

a) Ca(OCl)$_2$
b) pTsOH

43

+

44

Fig. 2.24. Formation of chlorinated by-product.

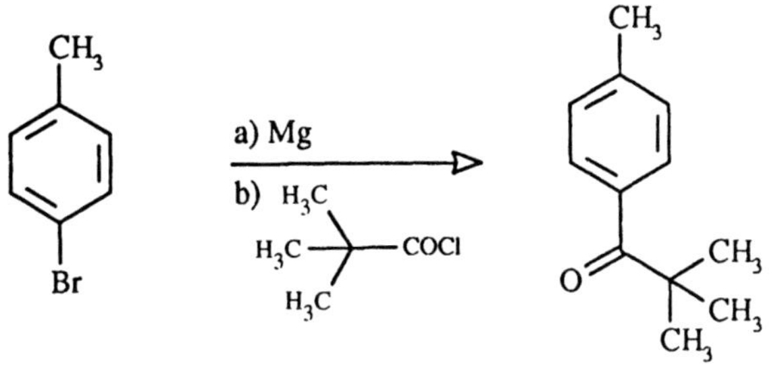

Fig. 2.25. Synthesis of SDZ 50-283.

2.11. EXAMPLE 11

Drug candidate **45** (Fig. 2.25) was prepared in one step from *p*-bromotoluene **46**, but the synthesis was plagued by the presence of an impurity, *p*-tolyl pivalate **47**, which co-distilled with the product, causing it to fail specifications (21).

An interesting second by-product was crystallized from the distillation residue and identified as di-*p*-tolyl sulfide **48** (Fig. 2.26). Its origin was never discovered but rubber septa were suspected. On the other hand, it was shown experimentally (reaction was run without a nitrogen atmosphere) that the ester impurity **47** arose from the reaction of the Grignard reagent with oxygen. This tolyl Grignard was so sensitive to air that only painstaking exclusion of air (including additional purification of nitrogen by an oxygen-removing cartridge) avoided the formation of the ester by-product. It turned out that IR spectroscopy was, surprisingly, a sensitive tool for detecting small amounts (2%) of the ester **47**; ^{1}H NMR, ^{13}C NMR, thin-layer chromatography (TLC), or gas chromatography (GC) could not compete. Adding the pivaloyl chloride to the Grignard reagent gave lower yields and purity of the product than adding the Grignard to pivaloyl chloride; an inverse addition was, therefore, recommended, although it involved a transfer of the extremely air-sensitive reagent. Furthermore, the reagent had to be kept warm or be sufficiently diluted to prevent solidification during the transfer and addition.

Because these issues could become worse on scale up, this complex problem with a simple molecule was attacked—and solved—by looking for other synthetic approaches. Engineering solutions can sometimes be found, but a chemical solution

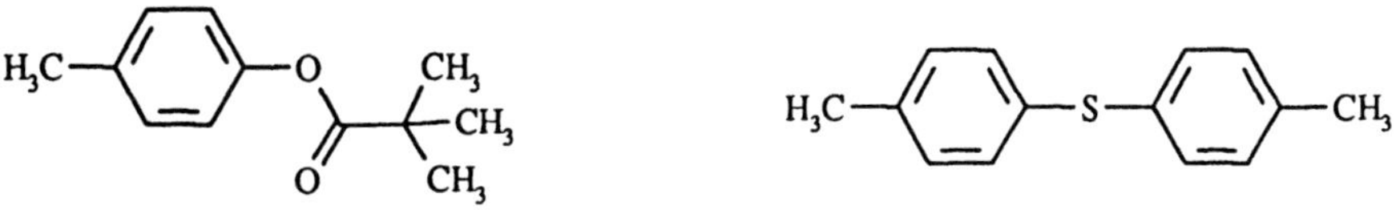

Fig. 2.26. Structures of two by-products.

was faster and cheaper in this case. When *t*-butyllithium was allowed to react with toluoyl chloride, no desired product could be detected. Similarly, Friedel–Crafts reactions failed: When aluminum chloride was added to a mixture of toluene and pivaloyl chloride or when pivaloyl chloride was added to toluene and aluminum chloride and the mixture heated, no desired product could be detected. Substituting zinc chloride for the Lewis acid and nitrobenzene for the solvent was not helpful. The reaction of *t*-butyl Grignard with *p*-toluoyl chloride produced an impure *t*-butyl-*p*-tolyl ketone. The solution finally came: The reaction of *p*-tolunitrile with *t*-butyl Grignard, followed by hydrolytic workup, gave a high yield of pure drug substance **45**.

2.12. EXAMPLE 12

The yield of transformation **49** → **50** was improved by using phase-transfer conditions (benzyltributylammonium bromide, toluene) (22). Rapid stirring was important for this two-phase reaction. Unfortunately, about 15% of a by-product formed that was identified by GC–MS as the alkylation product **51** (Fig. 2.27). This impurity could be entirely avoided by switching the solvent from toluene to dichloromethane (the opposite of Example 3). A mechanistic explanation for this solvent effect may be that dichloromethane is more acidic (i.e., contains some HCl) and thus suppresses benzylic anion formation. Regardless, the yield and purity were improved to 99 and 98%, respectively.

2.13. EXAMPLE 13

Diastereomers are common by-products in syntheses. The penultimate step (**52** → **53** in Fig. 2.28) of the convergent synthesis of fluvastatin (see Chapter 3) required the deprotection of the *syn* silyl ethers with tetra-*n*-butylammonium fluoride and acetic acid in tetrahydrofuran.

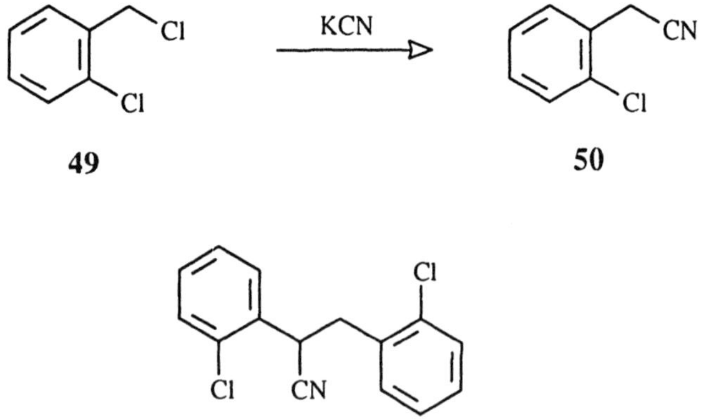

Fig. 2.27. By-product formation in a benzylic substitution reaction.

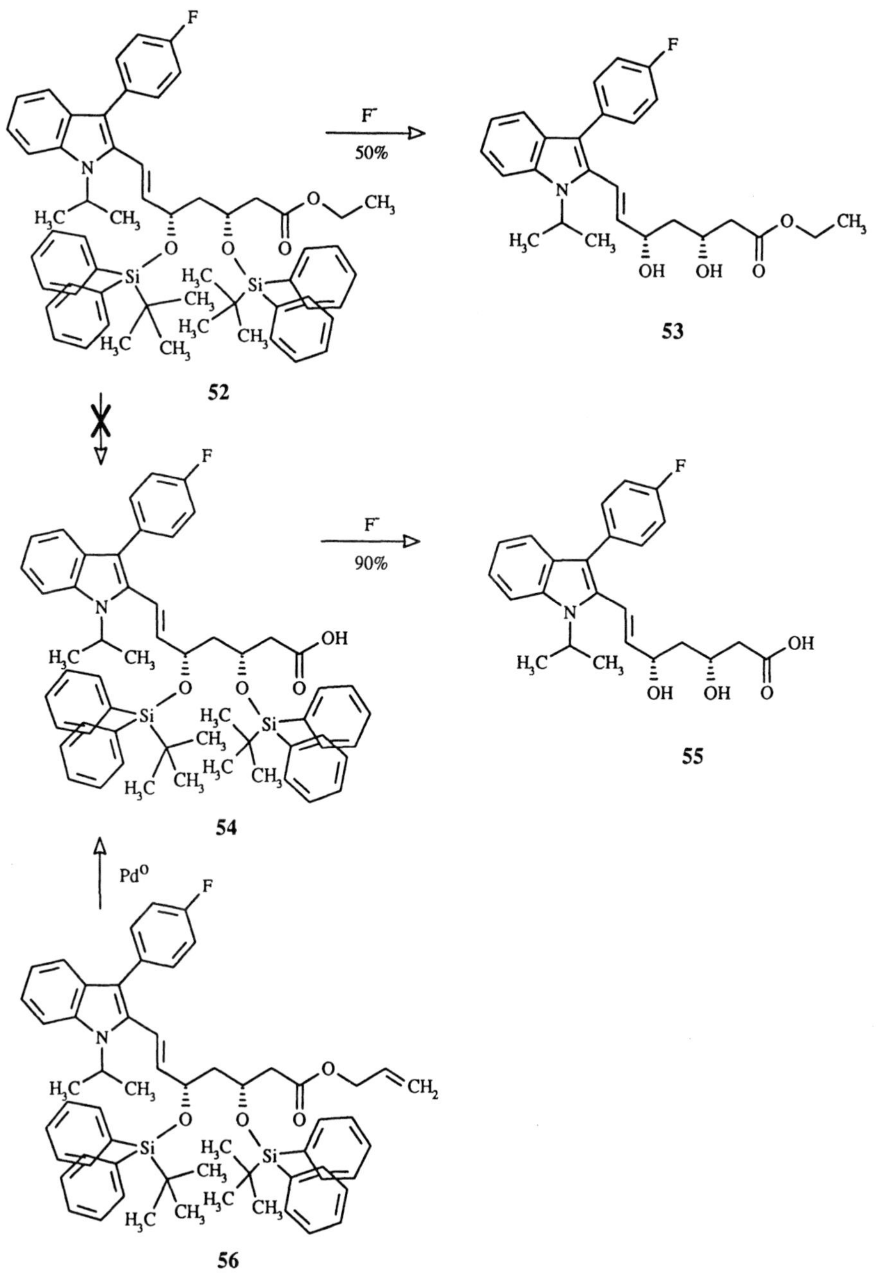

Fig. 2.28. Deprotection of silyl ethers.

Only a 50% yield of diol **53** was obtained in spite of (or because of) a 5-day-long reaction time required for the deprotection. The major by-products were olefinic elimination products and the 6-membered lactone, which isomerized and, on saponification, gave the undesirable *anti*-diastereomer of the drug substance (for the mechanism, see Fig. 3.29, Chapter 3). The desired reaction could be improved (90% yield in 2 days) by replacing the ester with a carboxylic acid (**54** → **55**) (10). The carboxylic acid had the following advantages: Because of general acid catalysis, it made the deprotection faster, minimizing side reactions; the acid was slower to form the lactone and, therefore, the *anti*-diastereomer; and it made the α-proton less acidic, minimizing the elimination side-reactions. Unfortunately, a switch to the carboxylic acid was not simple, as the ethyl ester could not be saponified (**52** → **54**). Instead, the carboxylic acid had to be introduced in a roundabout way, via an allyl ester **56**, which could be removed under neutral conditions with Pd^0; this ester, however, required a change early in the synthesis.

2.14. EXAMPLE 14

This is another example of a diasteromeric by-product. Drug candidate **57** (Fig. 2.29) is the carnitine ester of a thiophosphate. The thiophosphate is stereogenic, but the research synthesis produced a 1:1 mixture of diastereomers, which required separation by chromatography. That is to say, the reaction gave 50% of a diastereomeric by-product. By considering the reaction mechanism, the reaction could be

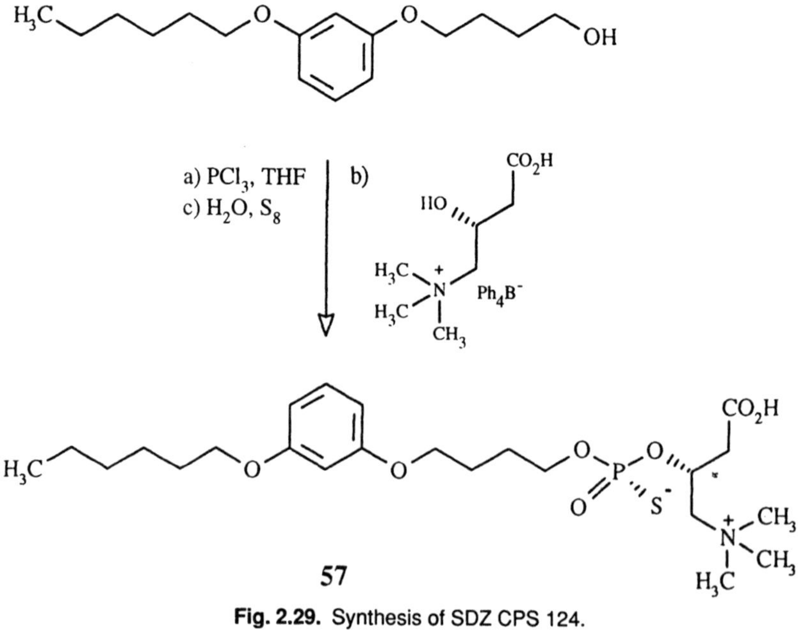

Fig. 2.29. Synthesis of SDZ CPS 124.

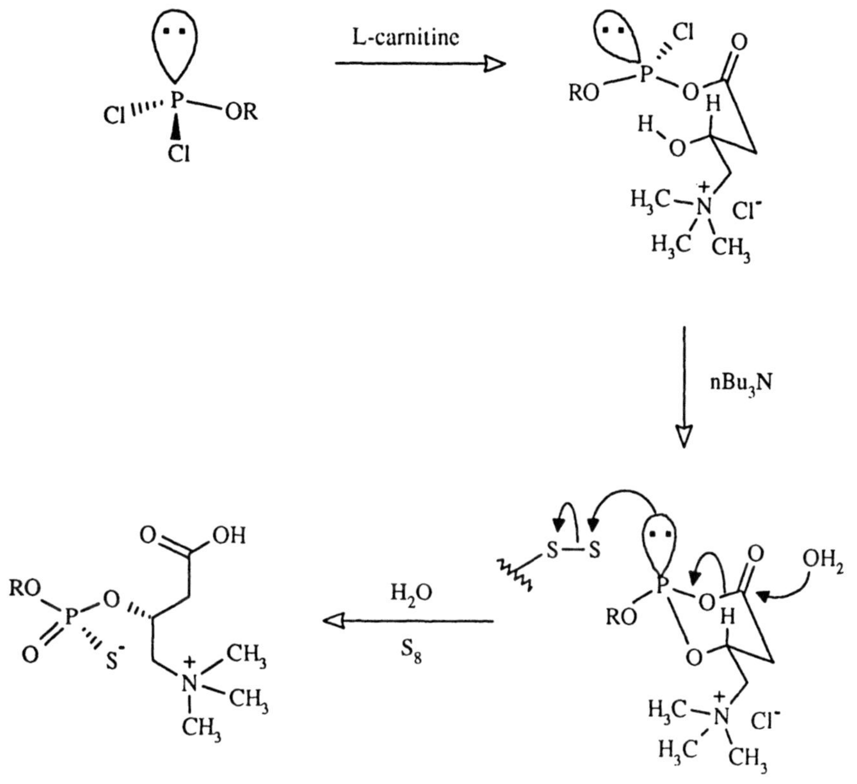

Fig. 2.30. Stereochemistry of thiophosphate formation.

improved (23). The selectivity is decided in the first reaction of this process step, the coupling of carnitine with the dichlorophosphite (Fig. 2.30), since apparently both subsequent reactions (the hydrolysis and the sulfurization) proceed with retention of stereochemistry on phosphorus.

The product is kinetically controlled: If the cyclic phosphite is aged, the ratio of diasteromers drops to $1:1$; and if it is heated, the undesired (thermodynamic) isomer predominates. This isomerization occurs presumably through ring opening and closing; some possible mechanisms are shown in Fig. 2.31.

The following reaction conditions gave the best results (see Fig. 2.30): Tributylamine was added to the dichlorophosphite in THF at $-15°C$, then carnitine tetraphenylborate was added at $0°C$ and stirred for no more than 30 min. After addition of water, sulfur was added. The sulfurization was slow (16 h at $20°C$), as it was solubility controlled.

Such considerations allowed the synthesis to be sufficiently redesigned and developed to improve the diastereoselectivity from $1:1$ to $3:1$, and the desired pure isomer could be crystallized from this mixture (24).

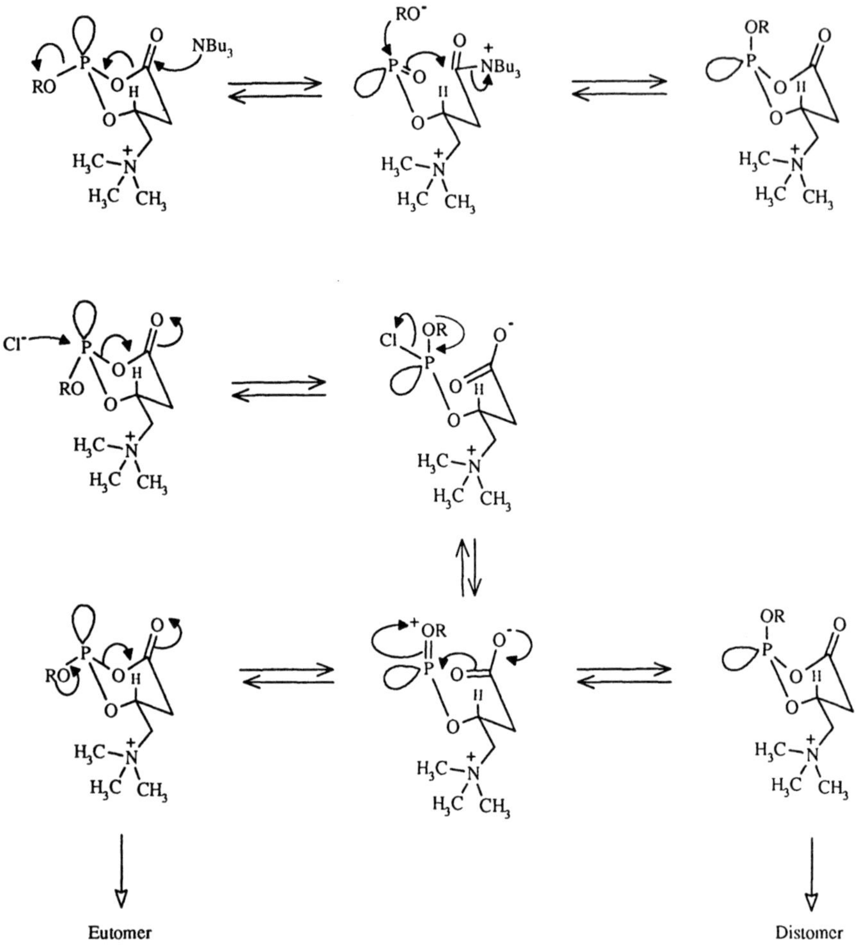

Fig. 2.31. Equilibration of diastereomers.

2.15. EXAMPLE 15

Drug substance SDZ WAG 994, **58,** underwent decomposition if heated at 100°C in 0.5% aqueous sodium hydroxide for 3 days (25). The major product of such a forced degradation was identified by LC–MS and NMR to be 9-cyclohexyladenine **59.** Even though this rearrangement was not a synthesis but a stability issue, it is mentioned here for its interesting mechanism (Fig. 2.32). We did synthesize **59,** however, as an analytical reference.

Back to the point, the major problem in the synthesis of **58** (Fig. 2.33) was the use of the protecting group, 1,3-dichloro-1,1,3,3-tetraisopropyldisiloxane. This reagent was expensive ($1,300/kg) and difficult to obtain on a bulk scale; its use

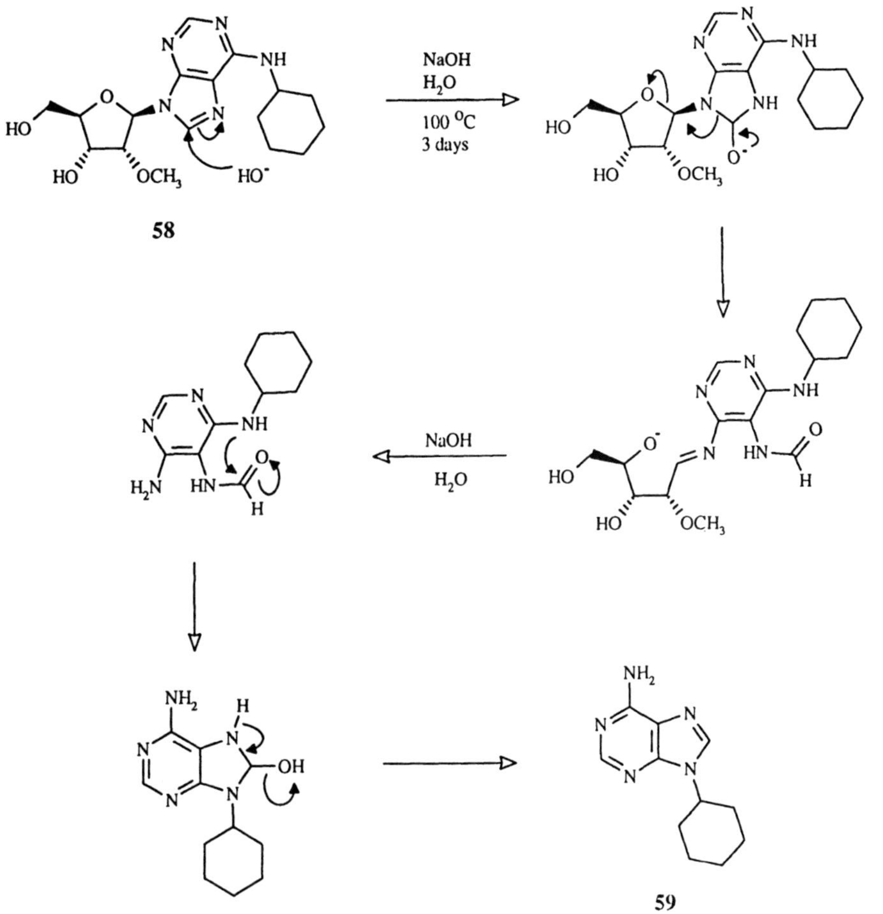

Fig. 2.32. Mechanism of apparent cyclohexyl migration.

added two steps to the synthesis, and it contaminated the product such that chromatography was needed in each of the last two steps. Furthermore, the protecting group formed another by-product (Fig. 2.34), which required a second chromatography for its separation.

To avoid these by-products and to shorten the synthesis, a most daring strategy was tried (26): using no protection at all (Fig. 2.35). The results were astonishingly good: The desired 2'-*O*-methylation **58** occurred to the extent of 68%. The surprisingly high regioselectivity can be explained by the fact that the 2'-hydroxyl group is the most acidic of the three because of the neighboring nitrogen (inductive effect); thus it is selectively deprotonated and methylated. The trimethylated and dimethylated products (23%) were removed by chromatography, and the 3'-methoxy by-product was removed by crystallization. The overall yield was the same as in the previous process, but with fewer operations and much lower cost.

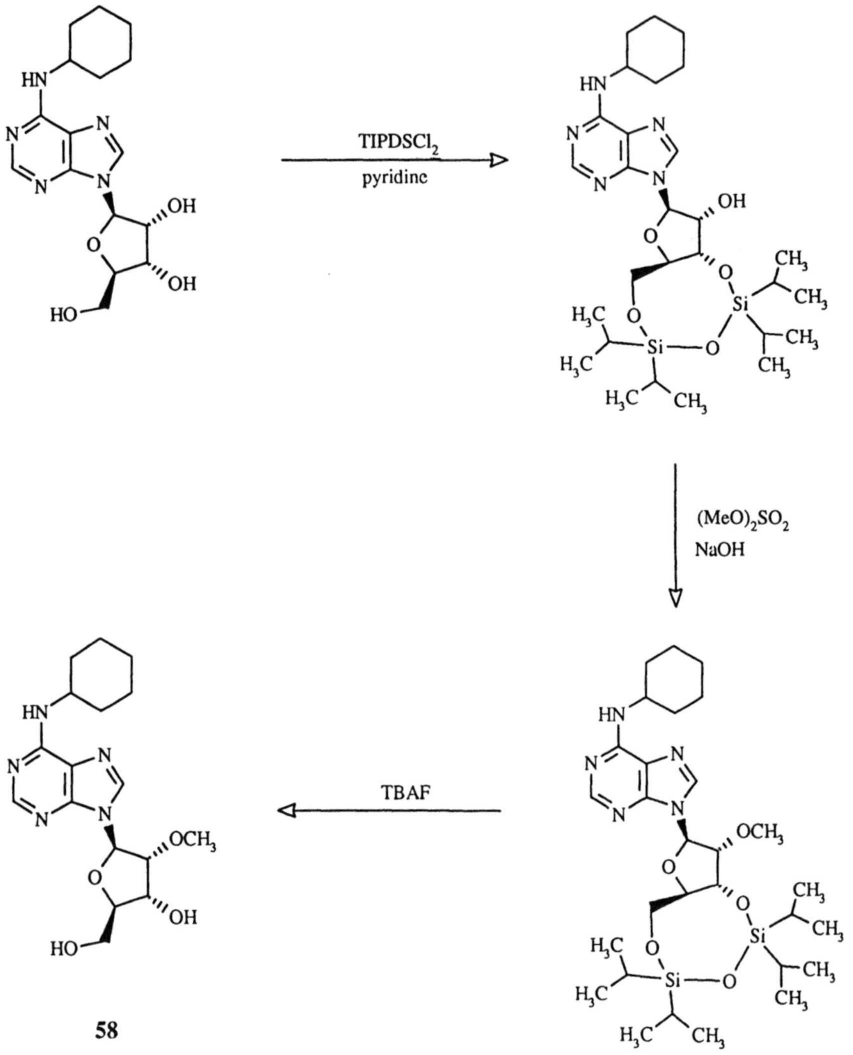

Fig. 2.33. Synthesis of SDZ WAG 994.

2.16. EXAMPLE 16

The last step (Fig. 2.36) of the synthesis of **60**, a platelet-activating factor inhibitor, required the most development (27); the original yield was low (34%) due to the formation of by-product **61**, which was, furthermore, difficult to separate. The solution was found by considering the mechanism of the reaction and of the by-product formation (28). The product of the desired condensation **62** is more acidic than the starting material **63**; thus **62** not only consumed half of the base and limited the yield to <50% but also, as enolate **64**, reacted with a second molecule of the ester, which led to the observed by-product **61**, thus further lowering the yield (Fig. 2.37).

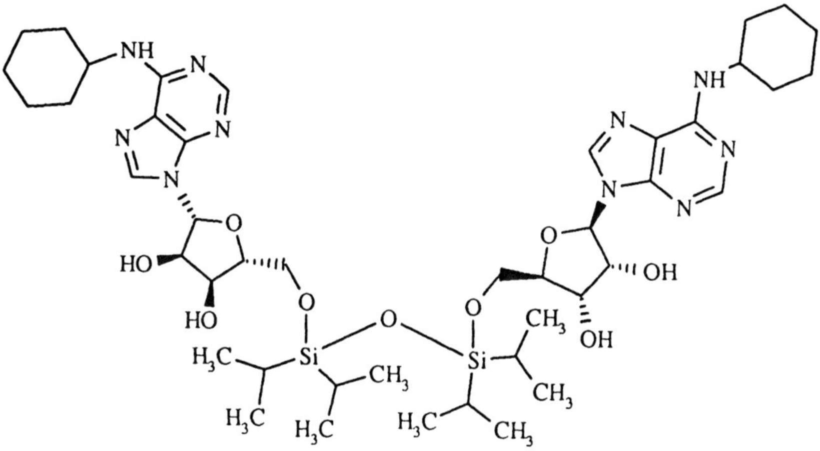

Fig. 2.34. Structure of major by-product.

It was, therefore, necessary to add—at the start—a third equivalent of *n*-butyllithium to regenerate the dianion and to complete the desired reaction **63** → **64**. It is noteworthy that, at low temperature, the extra *n*-butyllithium does not react with the ester and that the cyclization of **64** does not occur until the acidic workup. Thus the yield was improved to 57%, which was acceptable considering the many transformations that occurred in this step (see Fig. 2.36). Furthermore, the product no longer required chromatography for its purification.

Inorganic impurities also often interfere with drug substance specifications. At first, the residue-on-ignition test was outside the limits set for drug substance **60**. The problem was traced to Ca^{2+} and Mg^{2+} in the water used in the synthesis; deionized water must often be used to avoid the incorporation of these ions into the drug substance, especially if the desired compounds are chelators or are isolated from water.

2.17. EXAMPLE 17

Drug candidate **65** was prepared from the acid chloride **66** and the potassium salt of hydroxamate **67** in tetrahydrofuran (Fig. 2.38) (29). About 25% of by-product **68** (Fig. 2.39) was observed in the reaction.

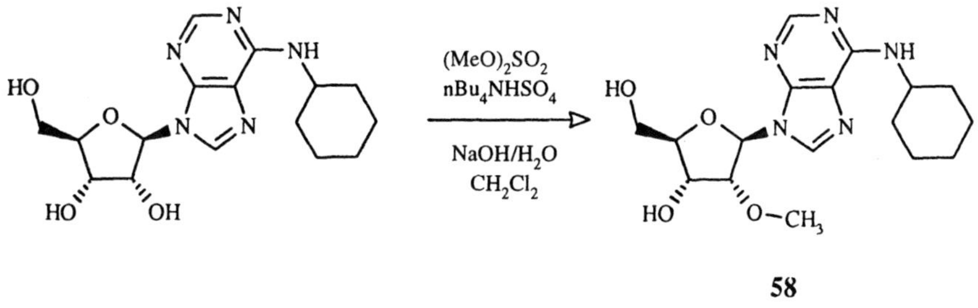

58

Fig. 2.35. Selective methylation without protection.

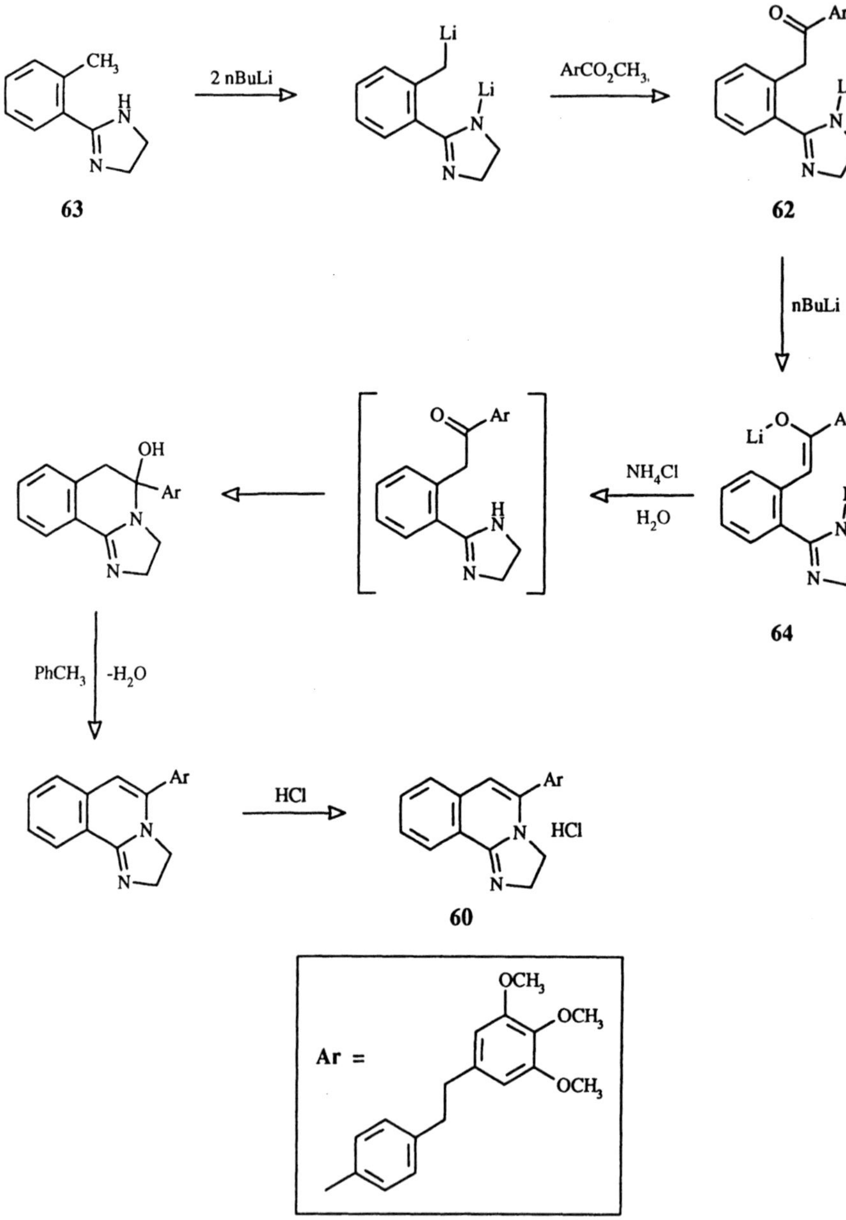

Fig. 2.36. Synthesis of SDZ HUL 412.

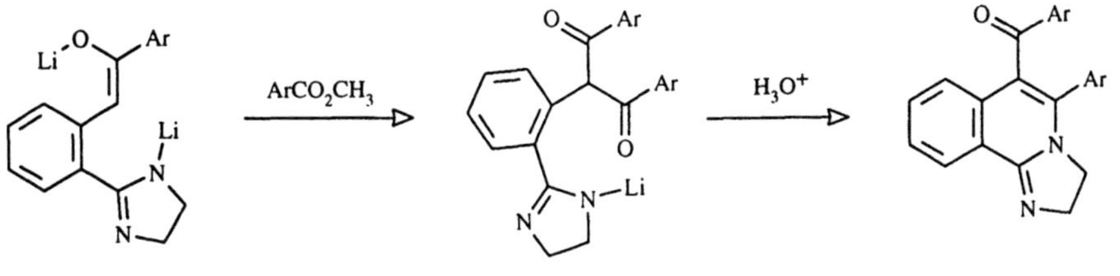

Fig. 2.37. Formation of undesirable by-product.

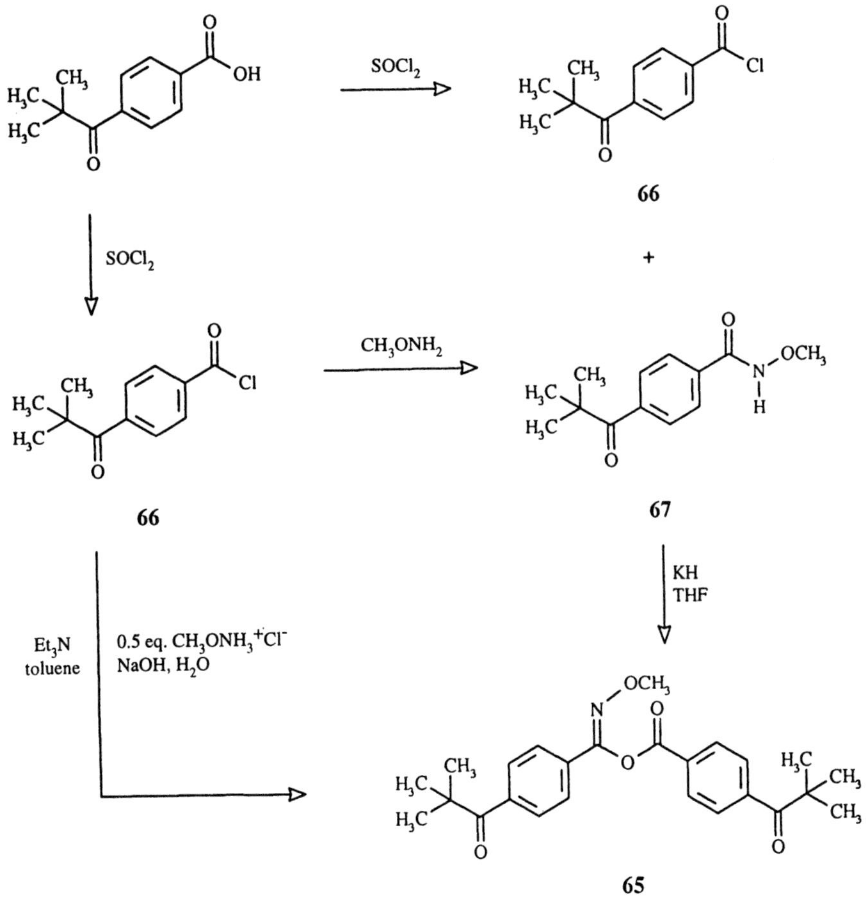

Fig. 2.38. Synthesis of SDZ FOX 988.

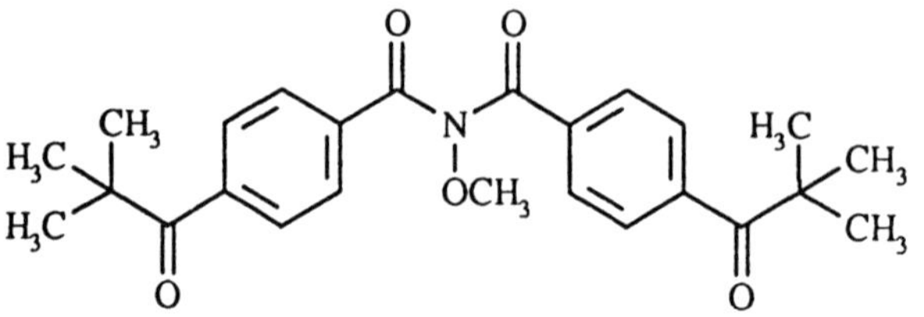

Fig. 2.39. Structure of the major by-product.

This side reaction seems to be the result of *N*-acylation as opposed to the desired *O*-acylation. The solution to this problem was found in a harder base (triethylamine) and in a more protic environment (toluene–water) (30); thus when acid chloride **66** was allowed to react with 0.5 equiv of methoxylamine in the presence of triethylamine, less than 3% of the by-product **68** was observed in drug substance **65**. The rationalization is complicated by the fact that the by-product **68** could also arise from **65** by a 1,3-acyl migration. This rearrangement is possible, presumably, only with the *E*-isomer **69** (Fig. 2.40), and so it may actually be the initial *E*:*Z* ratio that determines the amount of the by-product. In either case, the new reaction conditions (**66** → **65**, see Fig. 2.38) favored **65** over **68** by 98:2.

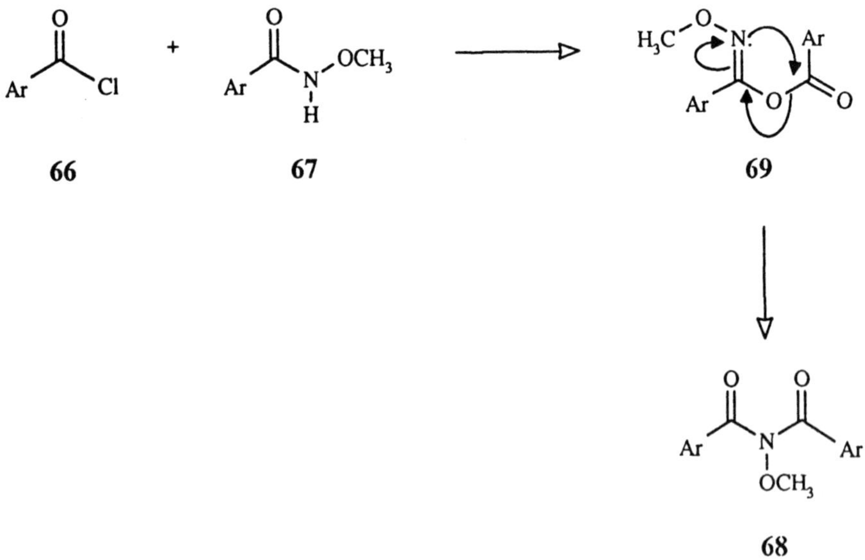

Fig. 2.40. Possible mechanism of by-product formation.

2.18. EXAMPLE 18

The previous example was a nonselective reaction, which, by definition, results in one or more by-products and is often a clue that one will encounter a low yield of the desired product. The opposite is not necessarily true, as illustrated by the following example (9). Although the Horner-Emmons reaction in Fig. 2.41 was highly diastereoselective (31), giving 100% of the *E*-olefin **70**, the yield was only 30%. Obviously, low yield can be a powerful indicator of by-products. To understand and improve the reaction, the mechanism of the conversion was studied. The condensation occurred readily in toluene at −78°C (*n*-butyllithium as the base) at which temperature, on quenching, one isolated the β-hydroxyphosphonate adduct. ^{31}P NMR of this material revealed that two isomers were present in equal amounts: 6,7-*anti*-**71** at 27.8 ppm, and 6,7-*syn*-**72** at 27.5 ppm (Fig. 2.42).

On heating to 20°C in the presence of the same base, the *syn*-isomer **72** eliminated to give the *E*-olefin, whereas the *anti*-isomer **71** in all cases remained and

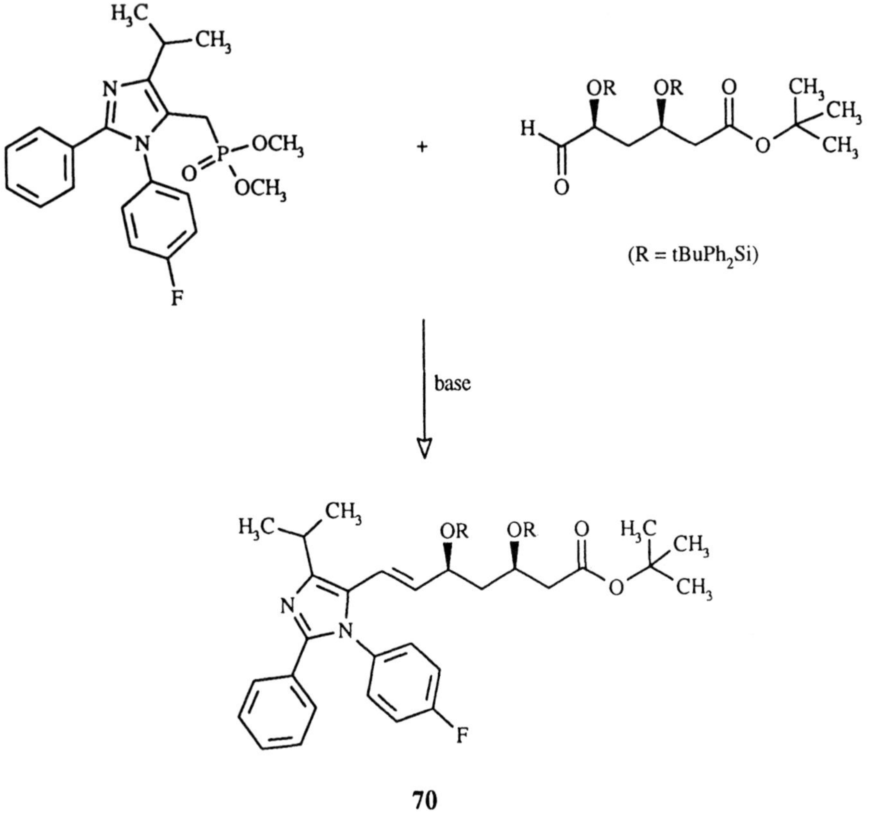

70

Fig. 2.41. Totally diastereoselective Horner–Emmons reaction.

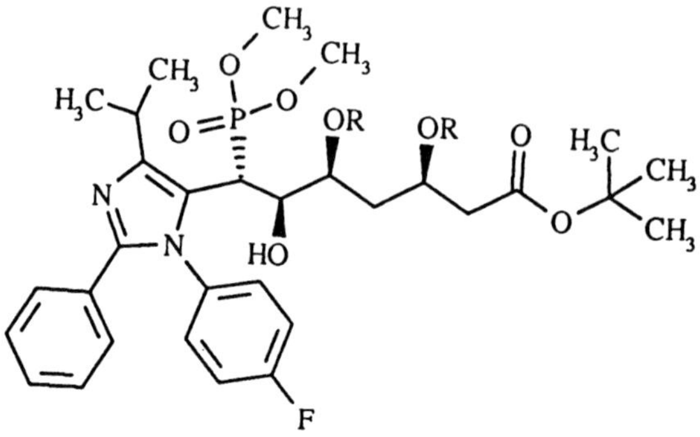

71

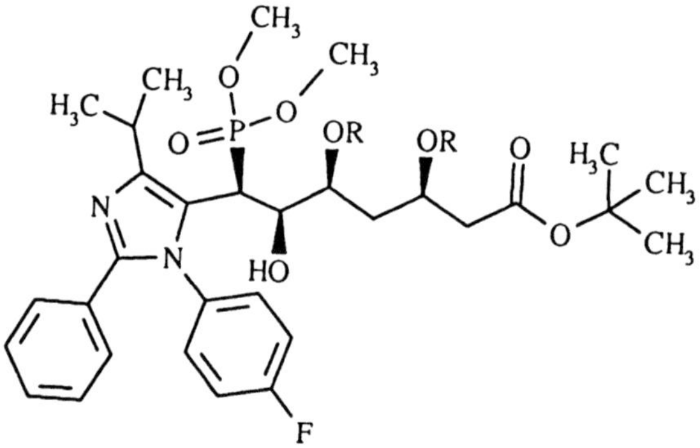

72

Fig. 2.42. Structures of *anti-* and *syn-β*-hydroxyphosphonates.

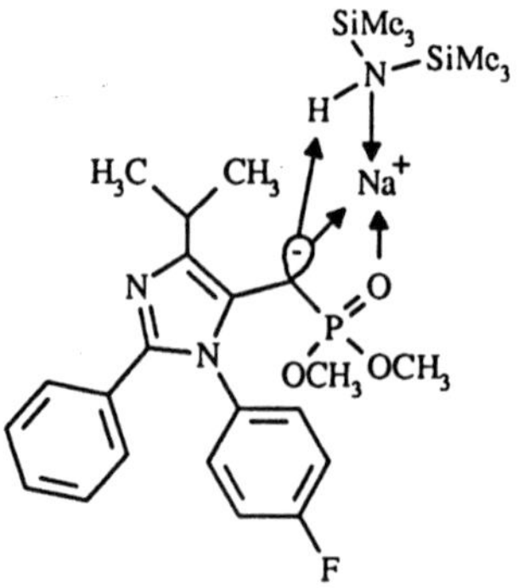

73 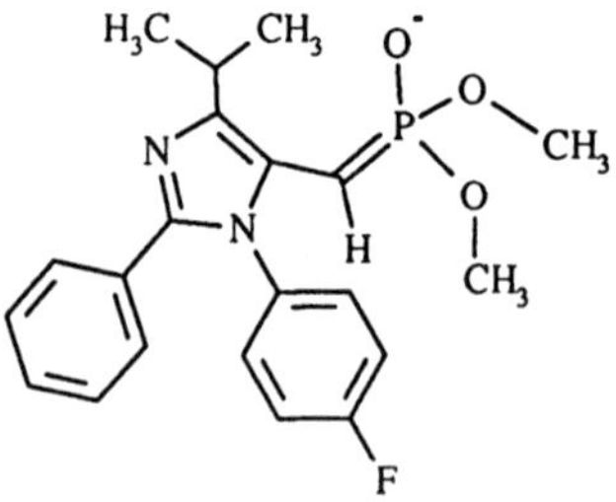   **74**

Fig. 2.43. Kinetic and thermodynamic phosphonates.

slowly decomposed to by-products, some of which were olefins that surprisingly retained the phosphonate. This observation explains why the reaction is so stereoselective: no Z-olefin is ever observed because the *anti*-adduct **71**, due to steric hindrance, cannot rotate and *syn*-eliminate. It also suggests two methods of improving the yield: either thermodynamically isomerize the *anti*-isomer into the *syn*-isomer (either by a reversible reaction or by deprotonation), or kinetically form predominantly the *syn*-adduct, which leads to the desired *E*-olefin. The first mechanism, thermodynamic equilibration through a reversible reaction, was excluded with a cross-over experiment; no incorporation of another, more reactive phosphonate was observed when it was added to the β-hydroxyphosphonate **71** and/or **72** under the same reaction conditions. The second mechanism à la Schlosser, isomerization by deprotonation α to the phosphonate, was excluded when the addition of a base did not improve the yield above 50%. This left the strategy for the formation of the *syn*-isomer by kinetic control. It was effected by a *quick* addition of the aldehyde and the use of sodium hexamethyldisilazide as the base. This improved the yield of the reaction to about 80%. The key intermediate may be the phosphonate C-anion **73** (Fig. 2.43), favored by amide bases, which slowly equilibrates with other bases (butyllithium or even the oxyphosphonate adduct) to enolate **74**. Rapid addition of the aldehyde perhaps traps the phosphonate anion **73** and gives rise to the desired *syn*-hydroxyphosphonate **71** (and *E*-olefin **70**) before the enolate equilibrium can be established that produces both *syn*-adducts and *anti*-adducts. Such kinetically controlled reactions can often be upscaled with continuous-flow reactors.

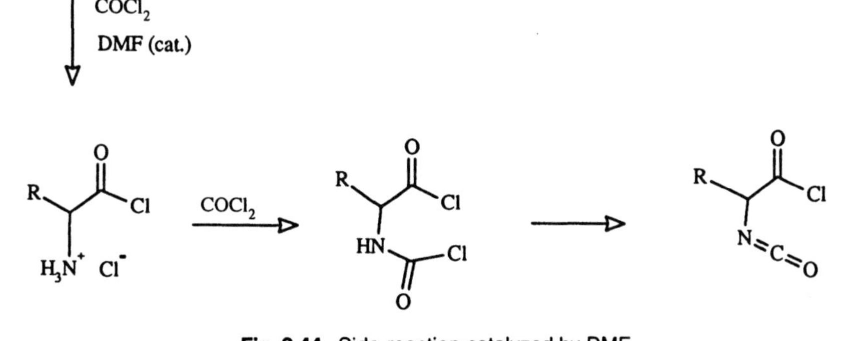

Fig. 2.44. Side-reaction catalyzed by DMF.

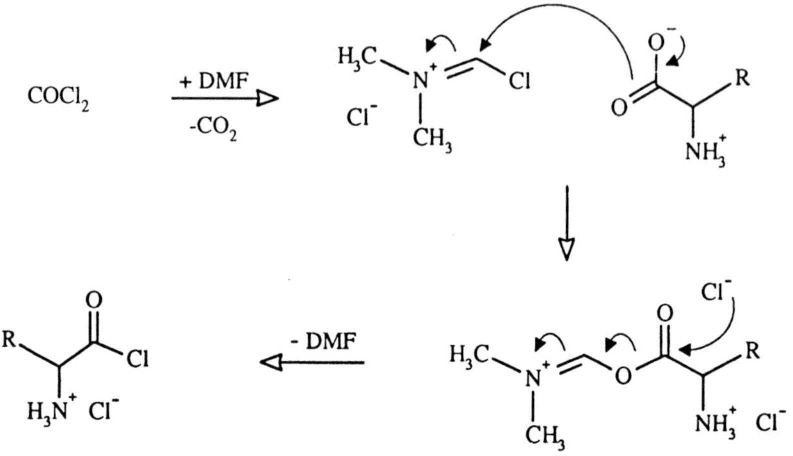

Fig. 2.45. Mechanism of acid chloride formation.

2.19. EXAMPLE 19

Sometimes small amounts of contaminants can catalytically alter the reaction pathway. For example, trace quantities of N,N-dimethylformamide present in the reaction of an amino acid with phosgene can result in the formation of an isocyanate **75** instead of the desired N-carboxyanhydride **76** (Fig. 2.44) (32). The probable mechanism is shown in Fig. 2.45.

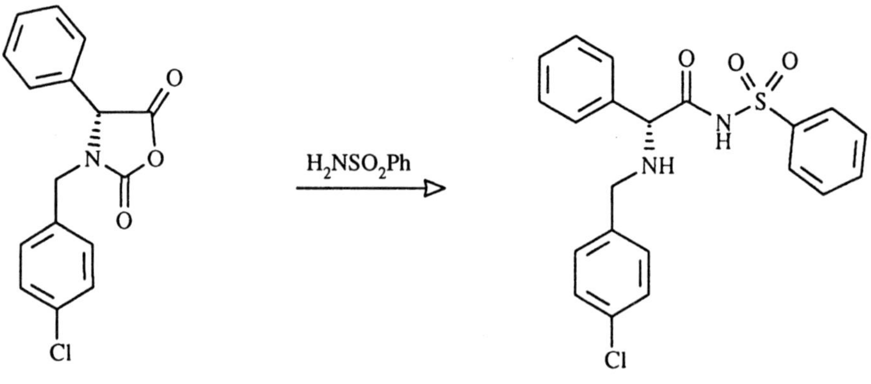

77   **78**

Fig. 2.46. Preventing racemization.

2.20. EXAMPLE 20

The undesired enantiomer can be considered an impurity in any synthesis of an enantiopure compound. Thus when phenylsulfonamide was added to a suspension of NaH in DMF, followed by the addition of *N*-carboxyanhydride **77** (obtained from enantiomerically pure (*R*)-phenylglycine), the product, drug candidate **78**, was found to be racemic (Fig. 2.46) (33). This disastrous result could be avoided by employing less basic reaction conditions, for example Et_2AlCl as the Lewis-acid catalyst (33) or, more conveniently, lithium phenylsulfonamide as the reagent (obtained by pre-mixing *n*-butyllithium and phenylsulfonamide) (34). In the latter case, a 77% yield of **78** was obtained with a chemical and optical purity of 99%, after crystallization.

BIBLIOGRAPHY

1. Eur. Pat. 539 326 (1993), "Bis-Phenylethane Derivatives, Their Preparation and Their Use for Treatment of Disease," P. Nussbaumer and A. Stütz (to Sandoz).
2. A. Kucerovy, T. Li, K. Prasad, O. Repič, and T. J. Blacklock, "An Efficient Large-Scale Synthesis of Methyl 5-[2-(2,5-Dimethoxyphenyl)ethyl]-2-hydroxybenzoate," *Org. Proc. Res. Dev.*, **1**, 287–293 (1997).
3. World Pat. WO 9618643 (1996), "New Tachykinin Antagonist Aminoacid Derivatives— Useful for Treatment of Pain, Inflammation, Emesis, Chronic or Obstructive Airway Disease, CNS Disorders and Allergic Diseases," S. Ko and C. Walpole (to Sandoz).
4. M. Prashad, and K. Prasad, personal communication, 1994.
5. D. Xu, K. Prasad, O. Repič, and T. J. Blacklock, "Ethyl Trifluoroacetate: A Powerful Reagent for Differentiating Amino Groups," *Tetrahedron Lett.*, **36**, 7357–7360 (1995).
6. U.S. Patent 4,287,355 (Sept. 1, 1981), "Carboxyl-(Phenyl or Tolyl)-Sulfonium Salts," F. G. Kathawala (to Sandoz).
7. Eur. Pat. 363 934 (1990), "Process for the Preparation of 7-Substituted Hept-6-enoic and Heptanoic Acids and Derivatives and Intermediates Thereof," K.-M. Chen, K. Prasad, G. Lee, O. Repič, P. Hess, and M. Crevoisier (to Sandoz).
8. U.S. Pat. 5,354,772 (Oct. 11, 1994), "Indole Analogs of Mevalonolactone and Deriva-tives Thereof," F. Kathawala (to Sandoz).
9. Eur. Pat. 244 364 (1987), "Production of 7-Substituted 3,5-Dihydroxy-6-heptene-1-oic Acid Derivatives, Including New Optically Pure Isomers, from Protected 6-Oxo-3,5-dihydroxyhexanoate Esters," K.-M. Chen, G. E. Hardtmann, K. Prasad, T. G. Lee, J. Linder, and S. Wattanasin (to Sandoz).
10. G. T. Lee, J. Linder, K.-M. Chen, K. Prasad, O. Repič, and G. E. Hardtmann, "A General Method for the Synthesis of *syn*-(*E*)-3,5-Dihydroxy-6-Heptenoates," *SynLett.*, **1990**, 508.
11. World Pat. WO 8603488 (1986), "New Indene Analogues of Mevalonolactone, Useful as Hypolipoproteinemia and Antiatherosclerotic Agents," F. Kathawala, S. Wattanasin (to Sandoz).
12. K. Prasad and R. Underwood, personal communication, 1988.
13. World Pat. WO 9003973 (1990), "Preparation of Pyrimidinyl-Substituted Hydroxyacids, Lactones, and Esters as Anticholesteremics and Hypolipemics," F. G. Kathawala (to Sandoz).

14. A. Kucerovy, P. G. Mattner, J. S. Hathaway, and O. Repič, "Improved Synthesis of Fluoroalkyl and Fluoroaryl Substituted 2-Aminopyrimidines," *Synth. Commun.*, **20**, 913–917 (1990).

15. DE 3,525,256 (1986), "New Naphthyl Analogues of Mevalonolactone, Useful as Cholesterol Biosynthesis Inhibitors," P. L. Anderson (to Sandoz).

16. Brit. Pat. 2,271,109 (1994), "Preparation of Substituted Benzylamine Derivatives," P. Nussbaumer and A. Stütz (to Sandoz).

17. S. Palermo and K.-M. Chen, personal communication, 1991.

18. A. Hofmann, *Die Mutterkornalkaloide,* F. Enke-Verlag, Stuttgart, 1964.

19. Brit. Pat. 2,170,407 (1986), "Pharmaceutical 9,10-Dihydrogenated Ergot Alkaloid-Containing Compositions." O. Zuger (to Sandoz).

20. M. Thiede and K.-M. Chen, personal communication, 1992.

21. U.S. Pat. 5,378,728 (Jan. 3, 1995), "Preparation of Benzoic Acid Derivatives as Antidiabetic Agents," J. Nadelson, W. R. Simpson, R. C. Anderson, J. S. Bajwa (to Sandoz).

22. U.S. Pat. 4,248,877 (Feb. 3, 1981), "(Hydroxypiperidine)acetic Acid Derivatives," E. Rissi and A. Ebnöther (to Sandoz).

23. U.S. Pat. 5,412,137 (May 2, 1995), "Process for Preparing Phosphinyloxy Propanaminium Inner Salt Derivatives," M. Prashad and K. Prasad (to Sandoz).

24. K. Prasad and D. Xu, personal communication, 1993.

25. Brit. Pat. 2,226,027 (1990), "New 6-Substituted, *eg.*, Alkenyl-2-O-Alkyl Adenosine Derivatives—Useful for Coronary Vasodilation, for Protecting Vascular Endothelium, for Lowering Blood Lipid Level, and Treating Increased Blood Pressure," F. Gadient (to Sandoz).

26. M. Prashad, K. Prasad, and O. Repič, "A Practical Synthesis of SDZ WAG 994 by Selective Methylation of N^6-Cyclohexyladenosine," *Syn. Commun.*, **26**, 3967–3977 (1996).

27. U.S. Pat. 4,992,428 (Feb. 12, 1991), "New 2,3-Dihydro-5-phenylimidazofuropyridine—Useful as Platelet Activating Factor Receptor Antagonists and Antitumor Agents," W. Houlihan and S. H. Cheon (to Sandoz).

28. J. C. Amedio Jr., U. B. Sunay, and O. Repič, "Metalated Aromatic Carboxylic Acids: Improved Synthesis of SDZ HUL 412," *Syn. Commun.*, **25**, 667–680 (1995).

29. Eur. Pat. 463 989 (1991), "*N*-Oxyimidic Acid Derivatives," K. Prasad, G. Lee, J. Nadelson, W. R. Simpson, and U. B. Sunay (to Sandoz).

30. U. B. Sunay, K. Talbot, K. Prasad, G. Lee, and L. Jones, "Synthesis of [$^{14}C_2$] SD2 FOX 988, A Hypoglycemic Agent," *J. Label. Compounds Radiopharm.*, **36**, 529–535 (1995).

31. B. E. Maryanoff and B. A. Duhl-Emswiler, "Trans Stereoselectivity in the Reaction of (4-Carboxybutylidene)triphenylphosphorane with Aromatic Aldehydes," *Tetrahedron Lett.*, **22**, 4185–4188 (1981).

32. T. J. Blacklock, R. Hirschmann, and D. F. Veber, "The Preparation and Use of *N*-Carboxyanhydrides and *N*-Thiocarboxyanhydrides for Peptide Bond Formation," in S. Udenfriend and J. Meienhofer, eds., *The Peptides,* Vol. 9, Academic Press, New York, 1987, p. 43.

33. T. Leenay and F. Kathawala, personal communication, 1989.

34. C.-P. Chen and K. Prasad, personal communication, 1990.

3

GOING ALL THE WAY

Perhaps the most satisfying moment for development chemists is the introduction of "their" new drug into the market, when years of labor are rewarded by knowing that their process produces a lifesaving therapy.

Herein is the story of one such project, Lescol, a drug that lowers serum cholesterol (1,2), from its discovery in the research laboratory through its introduction into the market (3). Chemical development was involved on and off throughout this period of about 10 years. During this time, exciting and innovative chemistry was performed (process research) and an ideal process was designed (process development). This chapter will describe this effort.

3.1. PROCESS RESEARCH

3.1.1. Discovery Synthesis

The process history of Lescol began with the structurally related compound SDZ 61-983. The chemical research and development department (R&D) received the experimental procedure from the discovery department (Fig. 3.1) for the synthesis of racemic compound **1**. A process review pointed out the following problems with the synthesis, keeping the criteria of an ideal process in mind:

1. The raw material p-fluorobenzyl bromide (**2**) is expensive and not available in bulk quantities (economy).
2. The preparation of diazonium salt **3** in the Fischer indole synthesis is dangerous (safety).

55

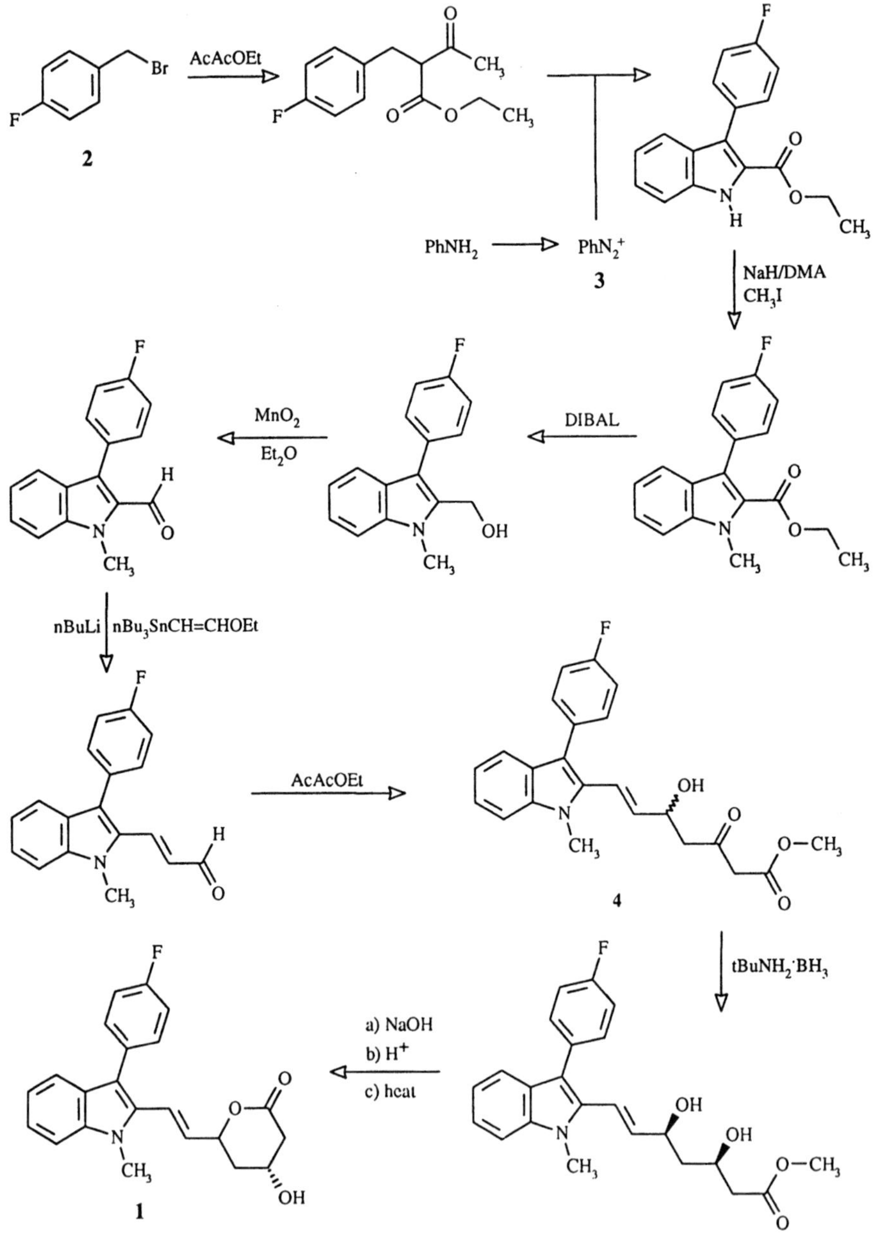

Fig. 3.1. Medicinal chemists' synthesis.

3. Manganese dioxide–diethyl ether oxidation is undesirable on a large scale (safety).

4. The reagent tri-*n*-butylstannylvinylethoxide is not available commercially and is difficult to prepare; Sn is an environmental hazard (economy, safety, ecology).

5. The reduction of the hydroxy-ketone **4** is not stereoselective; and a difficult chromatographic separation of the *cis/trans* lactones **1** is called for, yielding only 15% of the drug substance in this step (economy, specifications, plant fit, ecology).

6. The synthesis is long, comprising 11 synthetic steps (economy).

When faced with so many problems, it was prudent to look for a new, shorter synthesis. Looking at the molecular structure **1**, we see that any new synthesis design will present three challenges: a short synthesis of the indole, a stereoselective synthesis of the *E*-olefin, and a stereoselective synthesis of the *syn*-1,3-diol in the side chain. Each of these challenges was addressed in turn and solved.

3.1.2. New Synthesis of the Indole Portion

By switching from the Fischer to the Bischler indole synthesis, problems 1 and 2 were bypassed (Fig. 3.2). On the other hand, this approach gave indole **5**, which lacked a substituent in the 2-position. Since a 2-formylation of indoles had previously not been reported, we had to invent it (4). Several reagents were tried (Table 3.1).

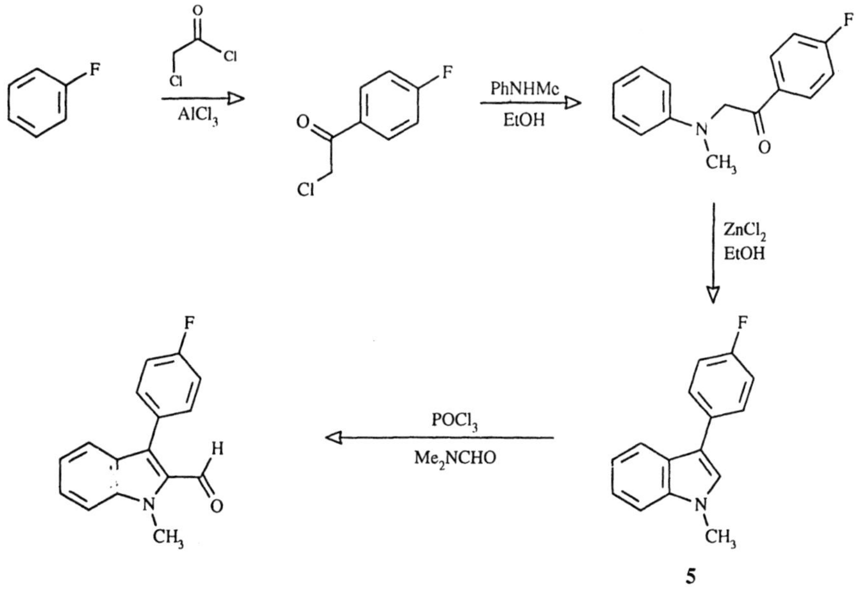

Fig. 3.2. Bischler indole synthesis.

Table 3.1. Formylation of Indole

Reagents	Yield, %
$SnCl_4/Cl_2CHOCH_3$	52
$AlCl_3/Cl_2CHOCH_3$	1
$TiCl_4/Cl_2CHOCH_3$	36
$POCl_3/(CH_3)_2NCHO$	50

The latter conditions resulted in the cleanest conversion to the desired product, and only a small amount of the undesired formylation at positions 5 and 7 was observed. This selectivity was important, as otherwise purification would become a new problem that the plant might not readily handle. This synthesis also eliminated problem 3. Regarding the moderate yield: We often consider a reaction successful if a yield of at least 50% is reached, with the assumption (based on an educated guess in each specific case) that the reaction can be further optimized in later stages of process development.

Two alternate synthetic approaches to the indole, based on *ortho*-lithiation, failed (Fig. 3.3). In the second case, lithiation occurred in the undesired position *ortho* to the fluorine; we exploited this observation successfully later in a radiosynthesis when we labeled a similar drug substance with a tritium atom for drug metabolism studies (see Fig. 7.10).

Our discovery of the 2-formylation process prompted research chemists to extend the idea and to attach a three-atom side chain, using *N,N*-dimethyl-3-aminoacrolein as the reagent (Fig. 3.4). This strategy further shortened the synthe-

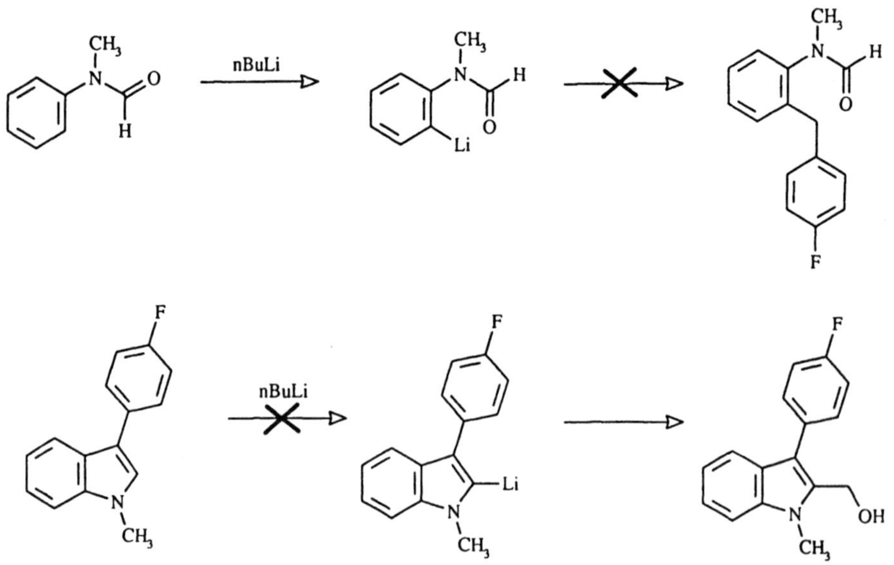

Fig. 3.3. Unsuccessful syntheses of indole portion.

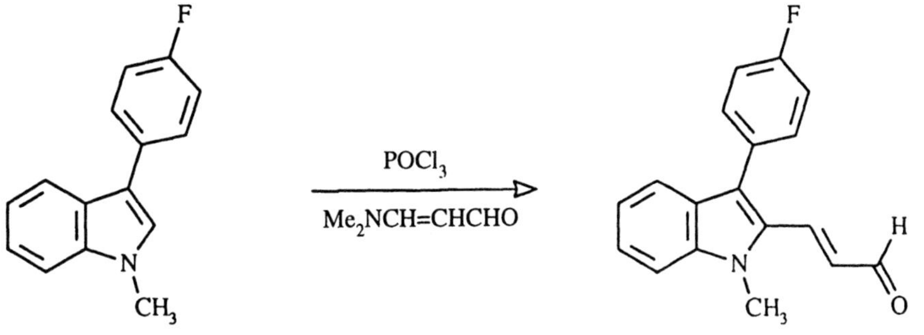

Fig. 3.4. Vinylogous formylation of indole.

sis by another step and avoided problems 4 and 5. Though this reagent was preferred to the tri-*n*-butylstannylvinylethoxide, it was also not readily available in bulk quantities and is toxic. The problem with its availability and synthesis will be mentioned again later in this chapter.

At this point, we were left with only one, but major, problem unresolved: the difficult separation of the diastereoisomers. To minimize the risk of failure, problem 5 was investigated and solved from three directions in parallel: a new, totally stereoselective synthesis was designed (see Section 3.1.3), a separation–purification method was designed that required no chromatography (see Section 3.2.2), and the reduction step was optimized until it was 99% stereoselective (see Section 3.3.5).

3.1.3. The First Stereoselective Synthesis of the Side Chain

So far we have seen the seven-carbon side chain built by attaching either $C_1 + C_2 + C_4$ or $C_3 + C_4$ carbon atoms to the indole; these two synthetic strategies we referred to as linear. A third way is possible ($C_1 + C_6$), which we called *convergent,* because it combined two larger molecules near the end of the synthesis (Fig. 3.5):

Path A (short formylation)	Indole + C + CC + CCCC
Path B (long formylation)	Indole + CCC + CCCC
Path C (convergent)	Indole + C + CCCCCC

Specifically, we designed the six-carbon synthon **12** for the side chain, to be attached in a Wittig coupling reaction to a properly functionalized indole (Fig. 3.6). We named this new synthon Prasad aldehyde (or Prasal) after one of our chemists.

The stereochemistry of the olefin was easy to obtain as synthetic methods are known that produce selectively *E*-olefins (but see Section 3.2.6). Of the stereochemical problems, the *syn* stereochemistry of the 1,3-diol was the most difficult to establish, as no known method was entirely stereoselective. Even a relatively high ratio of isomers of, let us say, 9:1 was not good enough, as this would still require elaborate operations for their separation.

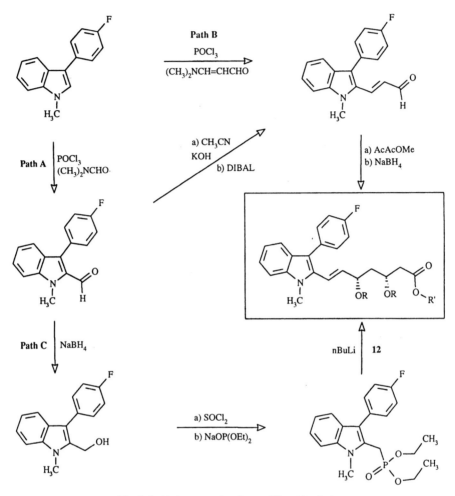

Fig. 3.5. Various constructions of the side chain.

The essential feature of the synthesis of **12** that Prasad (5) designed was its complete stereoselectivity (Fig. 3.7). The stereochemistry of the diol was created on a six-membered ring, which is easier to accomplish than it would be using a freely rotating acyclic system; the third functional group was used to open the ring and to adjust the oxidation state at both ends of the chain.

Specifically, phloroglucinol **6** was reduced with hydrogen on Rh/C to obtain a mixture of *cis,cis*-cyclohexane-1,3,5-triol and *cis,trans*-cyclohexane-1,3,5-triol (**7**). The *cis,cis*-triol could be separated by crystallization from the *cis,trans*-triol. The separation of isomers was not necessary, as it turned out later, because on protection with two equivalents of *t*-butyldiphenylsilyl chloride, only the *cis*-hydroxyl groups reacted to give **8**; the *trans*-hydroxyl happened to be axial and, therefore, more hindered. In other words, even the mixture of triols produced only the *cis*-

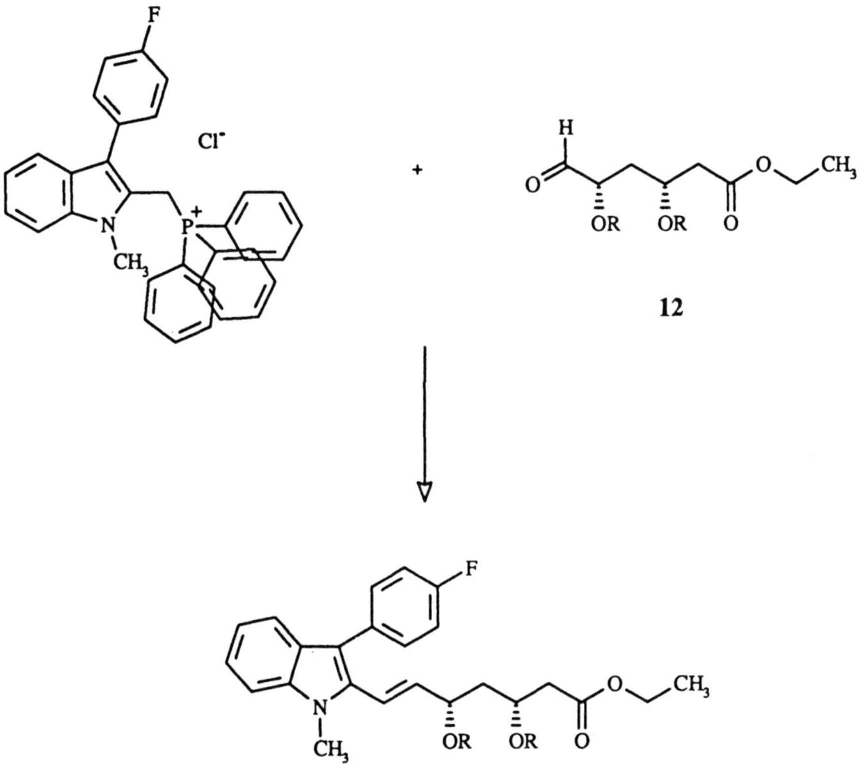

Fig. 3.6. Convergent synthetic route.

diprotected triols **8**, the *trans*-hydroxyl remaining unreacted. The stereochemistry of this third hydroxyl group did not matter, because in the next step it was oxidized to ketone **9** (with pyridinium chlorochromate) and further to lactone **10** (with *m*-chloroperbenzoic acid in a Baeyer–Villiger reaction).

All the atoms of the synthon were now in place. The seven-membered lactone was opened with ethanol in the presence of trifluoroacetic acid to give the alcohol ethyl ester **11** (6). The primary alcohol was oxidized (again with pyridinium chlorochromate) to aldehyde **12**, which was the oxidation state required for the subsequent Wittig reaction. Notice how the *cis*-diol in the cyclohexane became the *syn*-diol in the product.

A word is necessary about the unusual *t*-butyldiphenylsilyl protecting group. The large protecting group was used to prevent hydration of the aldehyde; it is known that α-alkoxy-aldehydes readily hydrate to form hemiacetals. A hydrate would, of course, affect the Wittig reaction. The presence of the protecting group prevented the hydration entirely. We believe the reason is not steric bulk (the synthon readily reacts in the Wittig reaction, after all) but its hydrophobicity. The bulky protecting groups make the environment around the aldehyde hydrophobic.

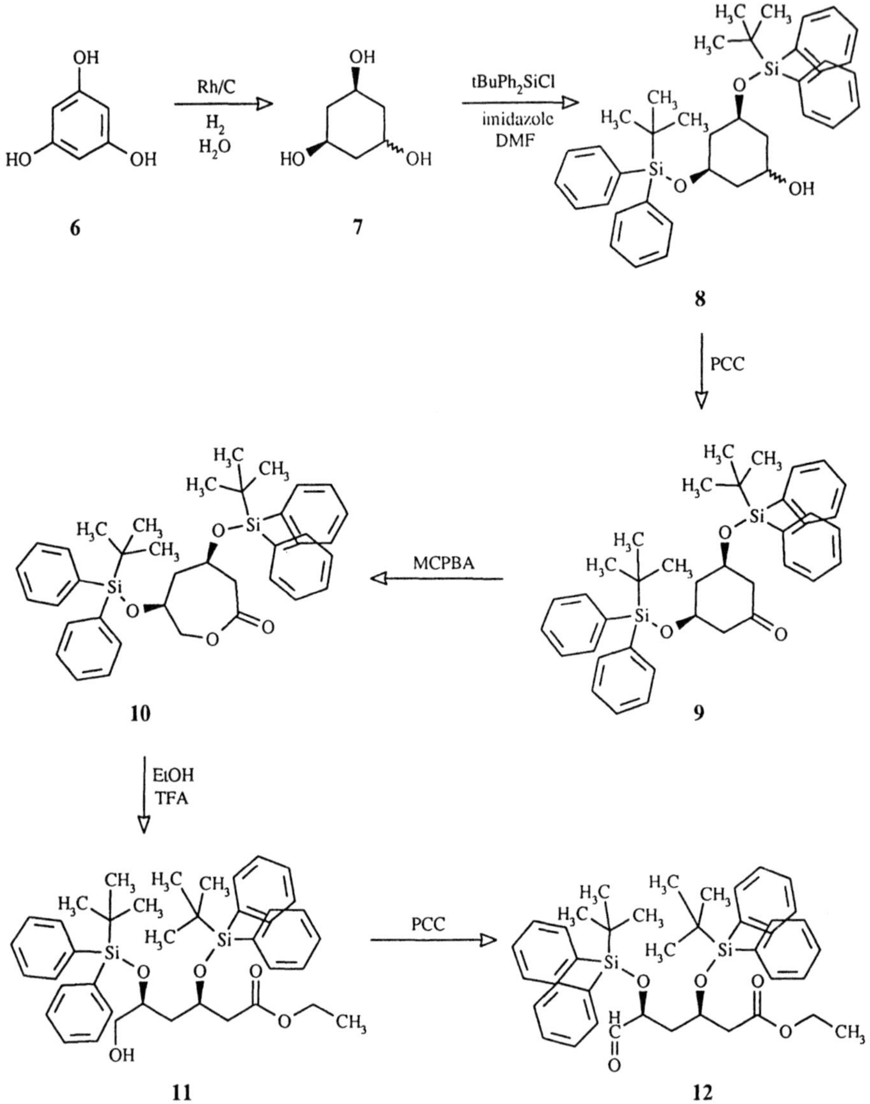

Fig. 3.7. Diastereoselective synthesis of side-chain synthon.

Aldehyde **12** was an excellent synthon for this and other related compounds (7), as it underwent a stereoselective Wittig coupling with many substrates. The synthesis was entirely stereoselective with respect to all three stereogenic functional groups (olefin and two carbinols). The stereoselectivity was the main strength of this new synthesis, and the objectives of process research were thus met.

To prepare the indole for the Wittig coupling, the "short" aldehyde **13** was reduced, instead of extending it to the conjugated aldehyde as before, and alcohol **14** was substituted with chloride to give **15** and then with triphenylphosphine to give **16** (Fig. 3.8).

3.1.4. Isomerizations

While the stereoselective synthesis was developing, several unsuccessful attempts were made to try to improve the stereoselectivity by *isomerizing* the *anti*-isomer into the desired *syn*-isomer. All these strategies involved forming six-membered heterocycles that could (but did not) isomerize in the 5-position (Fig. 3.9).

3.2. PROCESS DEVELOPMENT

In general, once a suitable synthetic pathway is established on a gram scale, one must turn attention to a plant process. Ideally, the thinking process has

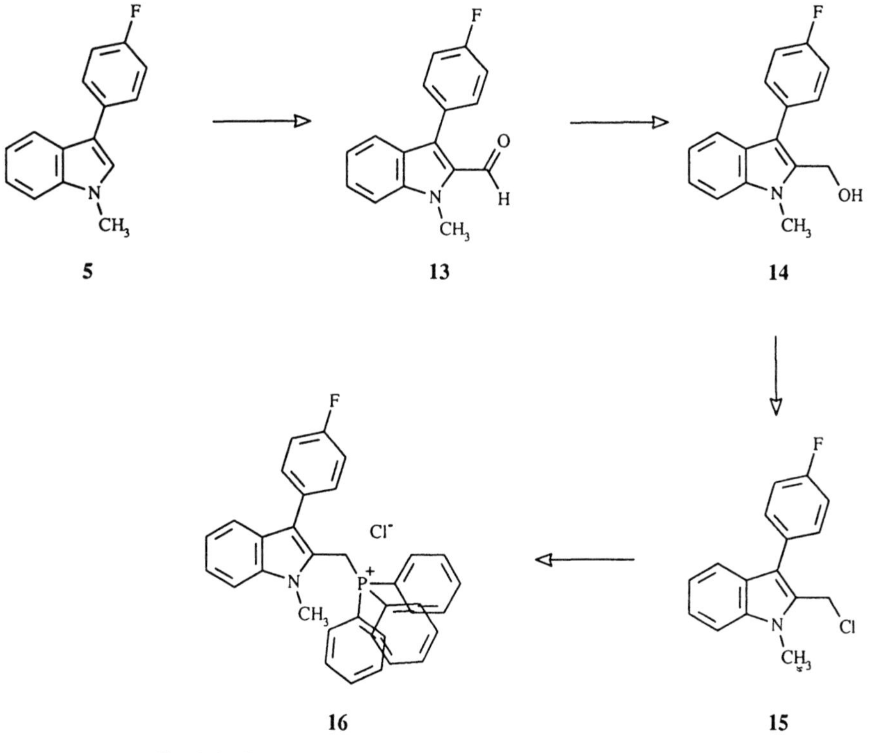

Fig. 3.8. Functionalization of indole for convergent synthesis.

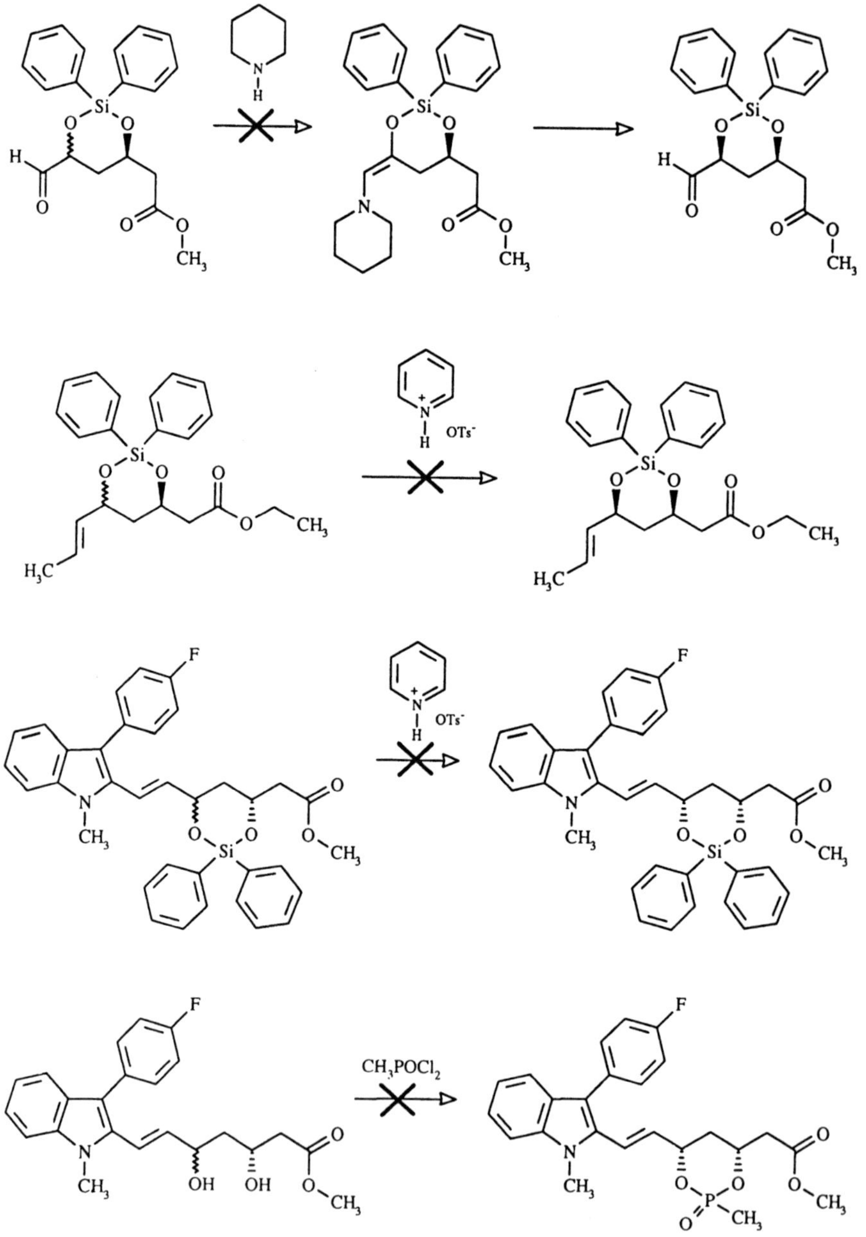

Fig. 3.9. Attempted isomerizations.

already begun during the process research phase. For each synthetic step, this entails:

- Defining physical parameters (reaction time, temperature, amount of solvent, etc.) so that they fit the plant.
- Defining chemical parameters (type of solvent, reagent, etc.) such that the identity, yield, stereochemistry, and so on are acceptable and reproducible.
- Finding acceptable workup and purification methods for the drug substance to pass specifications.
- Conducting a literature safety search and preparing samples for safety tests.
- Preparing samples of intermediates for analytical research.
- Conducting a raw-material search and cost estimate.
- Scaling the reaction up to a 5- or 12-L scale and writing a process based on this experiment.
- Performing a risk analysis (process safety, ecology, and industrial hygiene).

In this way, process development was completed through the phosphonium intermediate **16**.

3.2.1. Change to *N*-Isopropylindole

Project SDZ 61-938 was canceled and replaced with SDZ 62-320, **23**, the structure of which differed from SDZ 61-983 in two ways: the *N*-methyl group was replaced with *N*-isopropyl, and the lactone was replaced with the open-chain sodium carboxylate group (Fig. 3.10). Although the differences seemed minor on paper, process development had to be reinitiated since even minor changes can have far reaching consequences. For example, the amination of **17** required a 3-h reaction time with the *N*-methylaniline but a 10-h reaction time with *N*-isopropylaniline; the resulting product showed an exothermic decomposition at 50°C in the case of *N*-methyl, whereas the isopropyl case **18** showed an exothermic decomposition at 30°C; the short formylation had been optimized to an 86% yield with the *N*-methylindole but proceeded in only 59% yield with the *N*-isopropylindole **19**, probably due to steric crowding. Forcing the reaction and trying to increase the yield resulted in formylation at wrong positions. Even worse, since the lactone was no longer an intermediate, the separation of (now *syn/anti*) isomers of **20** was not possible even by chromatography.

The new synthesis contained the following weaknesses that had to be eliminated by process development:

1. *N*,*N*-dimethyl-3-aminoacrolein was not readily available; the single supplier delayed our work several times by missing delivery dates (deadlines).
2. The step leading to **20** did not meet several criteria for an acceptable process:
   - Triethylboron is spontaneously flammable in air (safety).

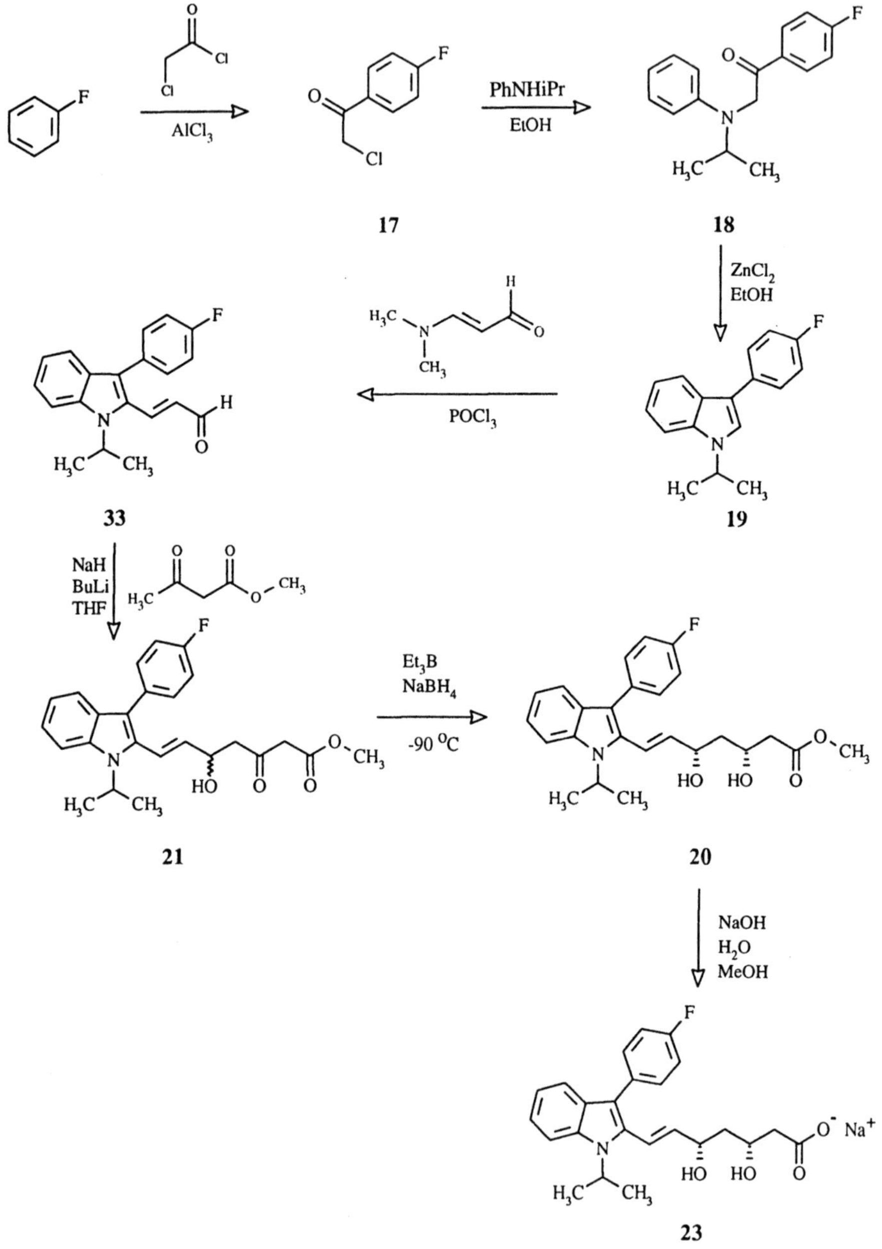

Fig. 3.10. First synthesis of fluvastatin.

- The reaction temperature of $-90°C$ was not available within the plant; later a liquid-nitrogen reactor was designed and installed (plant fit).
- The reaction was not stereoselective and resulted in a $80:20$ ratio of isomers that could be purified by ether extractions to no higher than a $91:9$ ratio (specifications).
- Intermediate **20** had to be converted to a lactone and purified by chromatography; a computer-controlled preparative HPLC was purchased and the intermediate was successfully purified, although about twenty 20-g injections were needed. Plant-scale chromatography equipment was subsequently purchased, but it was not an acceptable production method (plant fit, economy).
3. Finally, the drug substance was isolated by freeze-drying; plant-scale freeze-drying equipment was not then available in our plant (plant fit).

With these major obstacles facing us, parallel development of both linear and convergent syntheses was undertaken to maximize the probability of success.

3.2.2. Process Development of the Reduction Step

For reasons described in the previous section, the stereoselective reduction was the weakest step in the synthesis. We, therefore, concentrated on the development of this reduction step (Fig. 3.11). A detailed description of these efforts follows.

First, intermediate **21** was isolated and crystallized (previously it had been carried into the next reaction as crude). The isolation offered two advantages: The intermediate was more stable when crystalline than in solution; thus purer, more defined starting material could be made available for the study of the next reaction.

Once pure intermediate **21** was available, the stereochemistry of the reduction step was studied (8). The development was carried out, as usual, by varying key parameters (solvent, reagent, reaction temperature) and by observing their effect on the key result, here the stereoisomeric ratio of **20** (Table 3.2).

The best reagent was thus found to be $Et_3B/NaBH_4/THF$ or $Zn(BH_4)_2/Et_2O$; however, the stereoselectivity could not be reproducibly optimized beyond the $80:20$ ratio, even at $-90°C$. A nonchromatographic method for the separation of the diastereomeric *syn*-1,3-diol and *anti*-1,3-diol was, therefore, needed. After a literature search and some experimentation (9), we noticed that boric acid reacts with 1,3-diols to form crystalline borates and that they can be purified by recrystallization from isopropanol. Later it was determined that the actual intermediate was the isopropyl borate **22**, the isopropoxy group having been added by the solvent (Fig. 3.12). More important, every crystallization removed half of the *anti*-isomer. Thus, starting with a $80:20$ mixture, one required four crystallizations to obtain at least 98% pure *syn*-isomer. The boron was then removed with methanol, which also served to co-distill the resulting methyl borate.

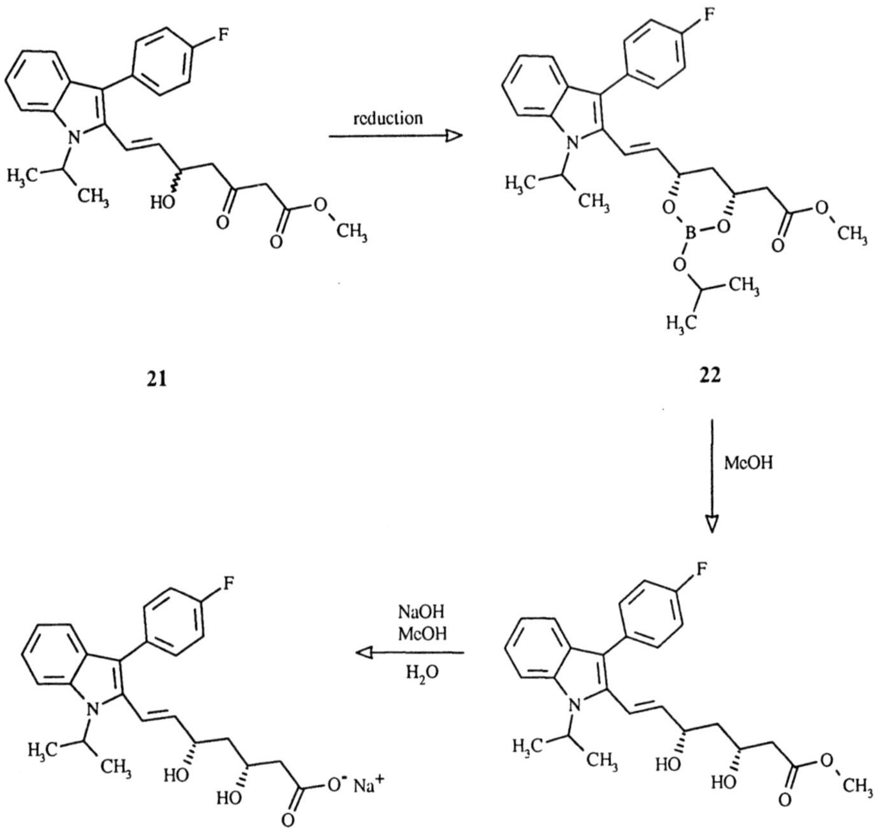

Fig. 3.11. Optimization of stereoselective reduction step.

With this breakthrough purification method in hand, we returned to the development of the reduction reaction conditions. The $Et_3B/NaBH_4$ method was discarded for the following reasons:

- Et_3B is toxic and pyrophoric (safety).
- Stereoselectivity was only 8:2 (specifications);
- Low temperatures ($-80°C$) were necessary for even this selectivity (plant fit).
- Long reaction times (>15 h) were necessary (economy).
- Air was needed to activate the Et_3B; this was an unsafe combination and difficult to measure (safety, plant fit).

The $Zn(BH_4)_2/Et_2O$ method was abandoned for the following reasons:

- $Zn(BH_4)_2$ could not be purchased and was difficult to prepare (reproducibility), and no analytical method was available for this reagent (specifications).

Table 3.2. Optimization of Disastereoselective Reduction

Solvent	Reducing Agent	Temperature, °C	Ratio, *syn* : *anti*
IPA	$NaBH_4$	0	50:50
MeOH	$NaBH_4$	−10	40:60
AcOEt	$NaBH_4$	−5	50:50
CH_3CN	$NaBH_4$	−10	40:60
IPA/AcOEt	$NaBH_4$	+10	50:50
IPA/AcOEt	$NaBH_4$	−10	60:40
IPA/AcOEt	$NaBH_4$	−20	60:40
IPA	$NaBH_4$	−50	60:40
THF	$NaBH_4/Et_3B$	−10[a]	80:20
THF	$NaBH_4/Et_3B$	−10[b]	70:30
THF	$Zn(BH_4)_2$	−45	75:25
ϕCH_3	$Zn(BH_4)_2$	−40	60:40
Et_2O	$Zn(BH_4)_2$	−20	70:30
Et_2O	$Zn(BH_4)_2$	−30	75:25
$tBuOCH_3$	$Zn(BH_4)_2$	−25	64:36
Et_2O	$Zn(BH_4)_2$	−45	80:20
THF	B_2H_6	−10	40:60
THF	$LiBHEt_3$	−60	50:50

[a]0.5 g.
[b]20 g.

· It worked well only in diethyl ether (safety).
· Stereoselectivity was only 8:2 and was not consistent (reproducibility).

Finally, for the sake of safety and reproducibility, $NaBH_4$ alone was used as the reagent and isopropanol and ethyl acetate as the solvents in the process, although it gave only a 6:4 mixture of diastereomeric diols. Five recrystallizations of borate **22** were required to obtain 98% pure *syn*-isomer; the yield of this step was consequently only 30% (Table 3.3).

At the time, this process was judged to be an acceptable compromise of economy and yield (which were poor) for the sake of safety, reproducibility, specifications, and plant fit (all of which were good) and was, therefore, used in the plant in the initial stages of development.

Once the desired stereochemical purity was obtained, the methanolysis to give **20** and the saponification to give the drug substance **23** were quickly developed, and process 1 was issued (see Fig. 3.11).

3.2.3. Optimization of Yield

Since five crystallizations of borate **22** were necessary to reach the desired stereochemical purity specification (and this operation nicely illustrates how costly it is to raise the purity of a substance to 98%), the yield of these crystallizations became important. There clearly exists a limit to how much one can compromise the yield, and thus economy, to reach desired purity.

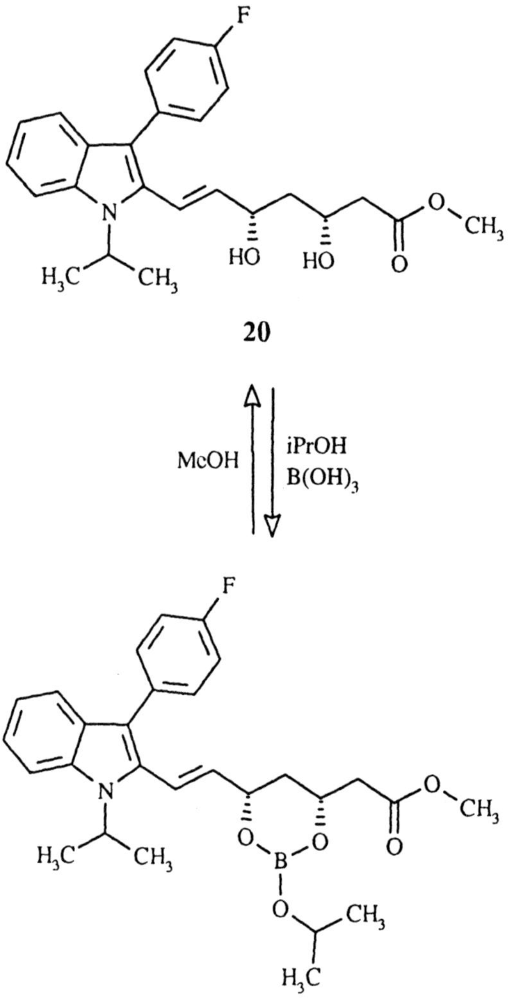

20

22

Fig. 3.12. Derivatization of diol for crystallization.

Table 3.3. Purification by Crystallization

Crystallization Number	*syn*	*anti*
0	60	40
1	75	25
2	89	11
3	94	6
4	97	3
5	98	2

Table 3.4. Recovery of Second Crops

Material	Amount, kg	*Syn* content, %	Yield, %
Crude	79	60	
First crop	15.8	99	20
Second crop	23	73	
Recrystallized	11.4	98	14
Total	27.2	>98	34

The yield of **22** was improved by making the following changes in the process workup:

· Hydrochloric acid was omitted.
· Crystallization time was extended from 3 to 18 h.
· Isopropanol : borate ratio was set at 15 : 1.

Furthermore, second crops were recrystallized similarly to recover more material (Table 3.4).

The total yield of >98% pure material was thus 34%; this low yield reflects the low stereoselectivity of the reaction (60%) and the consequent losses during purification operations, and this low yield will be addressed again with further process optimization (see Section 3.3.5).

3.2.4. Minimization of Boron Content

As mentioned earlier, the borate can be removed from the product with methanol. However, a careful analysis revealed that 2000 ppm of boron remained in the product (our specification for boron was 10 ppm). Several different treatments of intermediate **20** were tried to remove last traces of boron (Table 3.5).

Table 3.5. Methods for Removal of Boron

Method	Boron, ppm
One additional crystallization	577
2 × Methanolysis reaction time	1,000
20 × Methanolysis reaction time	691
Water wash of ethyl acetate solution	645
Crystallization from methanol	20,900
1 × Azeotropic methanol distillation	61
2 × Azeotropic methanol distillation	42
3 × Azeotropic methanol distillation	10
4 × Azeotropic methanol distillation	8
5 × Azeotropic methanol distillation	<5

These results confirmed that the impurity was methyl borate, which forms an azeotrope with methanol (54.6°C at 1 atm, in a ratio of 34:66 methanol:methyl borate) (10). The fact that so many distillations are needed to remove all of methyl borate may indicate that it exists in an equilibrium with diol **20** (Fig. 3.13). Fortunately, this elaborate purification treatment did not change the assay or the stereochemistry (*syn : anti* ratio) of the product, and this method for boron removal was successfully used in the plant to prepare the early batches of material.

The number of azeotropic distillations needed on large scale sometimes exceeded the number needed in the laboratory and was, furthermore, variable. The reason was presumably that the distillation in the plant could not remove as much solvent as in the laboratory, making each azeotropic distillation less efficient and more were needed. To avoid this time- and energy-consuming operation, we continued to look for other methods of boron removal.

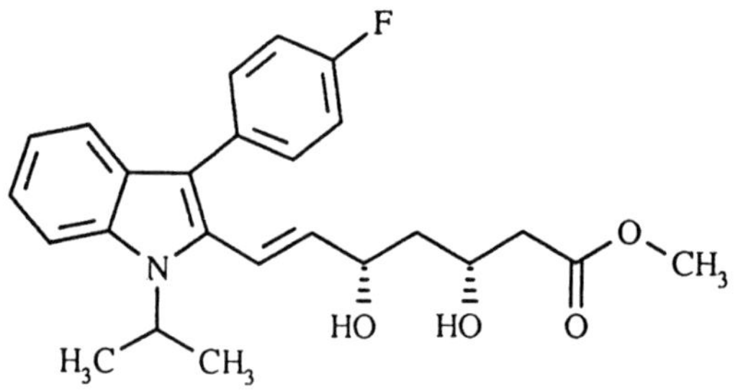

20

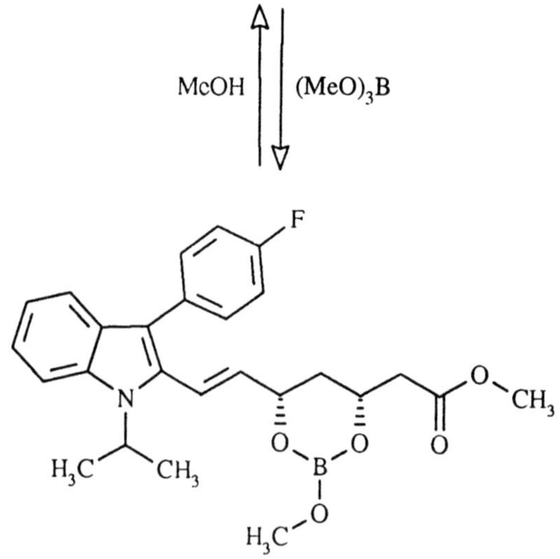

Fig. 3.13. Methyl borate–diol equilibrium.

Many materials were tried with little success; apparently **20** was a better chelating agent for boron than most other materials tried, as it extracted boron even from borosilicate glass! One dramatic exception was the ion exchange resin Amberlite IRA-743, which completely removed all boron from our samples (Table 3.6).

Though successful, this solution was never used by the plant because they preferred distillations to ion exchange chromatography and because later an easier solution to the boron specification problem was found (see Section 3.3.5).

3.2.5. Development of the Side-Chain Synthon Synthesis

As is clear from the above discussion, the linear synthesis was not stereoselective. Many purification operations were needed, which contributed to the low total yield. To improve the situation, we also developed a second approach, the stereoselective and convergent synthesis mentioned earlier, using the Prasal synthon **12** (see Fig. 3.7).

The preparation of **8** was complicated by the rapid conversion of the disilylated triol **8** to the trisilylated triol. The yield of the desired product **8** could be improved by using a higher content of *cis,trans*-triol in **7**, as the axial (*trans*) hydroxyl silylated slower. Accordingly, an optimization of the reaction **6 → 7** was attempted to maximize the *cis,trans*-triol and minimize the *cis,cis*-triol in **7**. The solvent, time, pressure, temperature, and catalyst were varied. The best conditions were found to be Rh on alumina in *t*-butanol, or Ru on carbon in water, both at 80°C for 2 h. However, only about a 1 : 1 ratio of stereoisomers **7** could be reached.

This low ratio lowers the yield of this step if isomers are separated by crystallization or of the next step if both isomers are used; but *it does not affect the stereoselectivity of the synthesis* as the epimerized (third) hydroxyl is anyhow eliminated by oxidation to **9**. This oxidation, as well as the analogous step leading to **12**, was improved physically and chemically by adding dried powdered molecular sieves to the reaction, which served as a catalyst, desiccant, and adsorbent. Although pyridinium chlorochromate is not an attractive reagent for the plant (carcinogen, environmental problems), no substitute could be found. The molecular sieves made its handling (filtration, isolation, and disposal) at least safer. A silica gel filtration step

Table 3.6. Methods for Removal of Boron

Reagent	Boron after Treatment, ppm
Polymer-bound triphenylphosphine	148
Amberlite IRA-743	0
Amberlite IRA-410	30
Chelex 100	70
Cellulose	80
Ultra-C charcoal	59
Neutral alumina	74
Celite	127
Sodium sulfate	109

was added to remove all traces of Cr from the product. To form lactone **10** via the Baeyer–Villiger reaction, *m*-chloroperbenzoic acid was used; although it is expensive, it is a relatively safe peracid with a relatively high decomposition temperature. This choice illustrates a compromise of economy for safety. Charcoal was an effective catalyst for the decomposition of peroxides. Beware, however, of *m*-chloroperbenzoic acid anhydride that can form during Baeyer–Villiger reactions: It is not detectable by peroxide paper tests.

Several kilograms of synthon **12** were prepared by our pilot plant for use in several cholesterol-biosynthesis inhibitor projects.

3.2.6. Development of the Convergent Synthesis

Next, aldehyde **24** had to be converted into a Wittig ylide (Fig. 3.14) that was to be used for the coupling with the synthon and the formation of the double bond. The transformation was accomplished by reducing **24** to alcohol **25** with sodium borohydride, converting it to chloride **26** with thionyl chloride, and finally to the triphenylphosphonium salt with triphenylphosphine. The sequence was improved by combining the last two reactions, thereby omitting the isolation of chloride **26**. The yield and purity of the chloride were also improved by inverting the addition of thionyl chloride: by adding the alcohol to thionyl chloride, the excess thionyl chloride prevented formation of the ether among the unreacted alcohol and the chloride product (see also Section 2.1.1). The ether would be especially noticeable on scaleup, since on a larger scale the reagent addition time lengthens and the problem worsens.

Once synthon **12** became available in kilogram quantities, the remaining steps of the convergent synthesis were developed. The Wittig coupling (Fig. 3.15) of **12** with **27** gave an 85:15 mixture of *E* and *Z* isomers. This ratio was optimized (to 100% *E*) by switching to the Horner–Emmons reaction, which uses a phosphonate rather than a phosphonium as the intermediate. *Thus all stereochemical problems in the synthesis were solved* (Fig. 3.16).

Two problems remained with the Horner–Emmons reaction, however. First, the yield was only about 50%; it was improved to 65% as follows:

- Triethyl phosphite was substituted with trimethyl phosphite, lowering the steric crowding in the reaction.
- The methyl ester was substituted with ethyl ester, suppressing undesired saponification during the reaction.
- LiCl was added for a salt effect.
- The phosphonate was added *to* butyllithium instead of the reverse; this order of addition prevented the methylation of the phosphonate anion by the unreacted methyl phosphonate, a side reaction that resulted in an extra methyl group on the olefin (see Fig. 2.13).
- Air was carefully excluded from the reaction mixture; air oxidation of the phosphonate anion produced an aldehyde that coupled with the remaining phosphonate to give a homodimer.

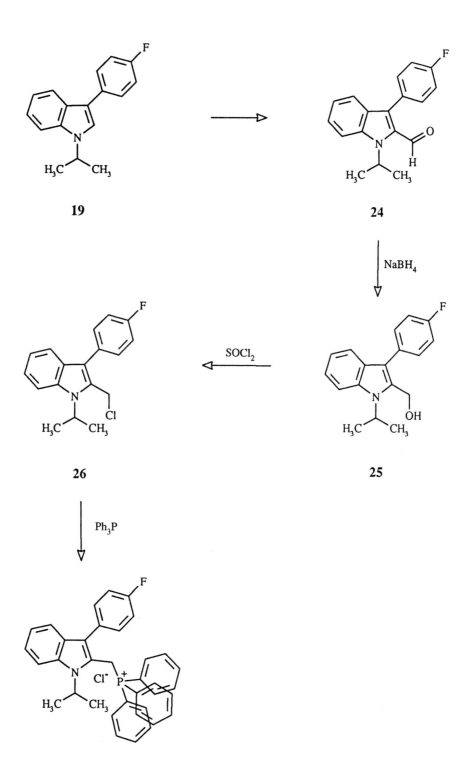

Fig. 3.14. Functionalization of indole for convergent synthesis.

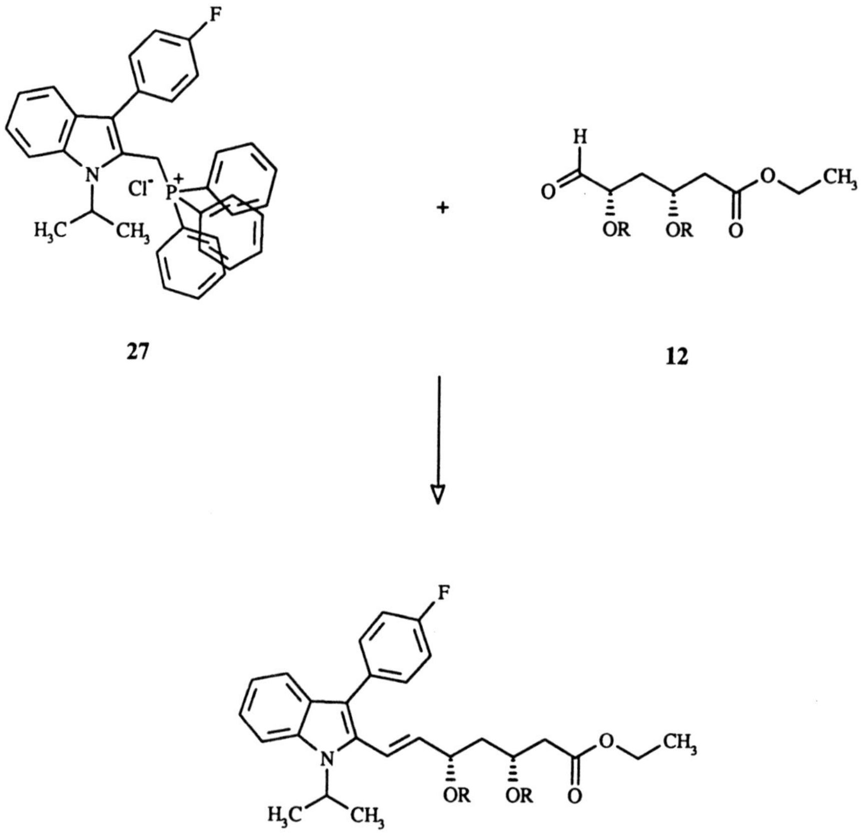

Fig. 3.15. Wittig coupling of side chain synthon.

The second problem was in the area of ecology and safety. The preparation of the phosphonate **28** (the Arbuzov reaction) produces methyl chloride; this side product is an alkylating agent and is, therefore, toxic. We solved this problem by switching the reagent from trimethyl phosphite to sodium dimethyl phosphonate, which produces environmentally more friendly sodium chloride as the only side product (Fig. 3.17).

3.2.7. Development of the Deprotection Step

The desilylation of **29** to **30** with fluoride also gave low yields (about 50%), and the initial reaction time was long (5 days at room temperature). If pushed too hard, some lactone formed, which tended to isomerize and give a few percent of the undesired isomer. It was minimized, but not eliminated, by switching from the methyl to the ethyl ester **29**.

On the supposition that the desilylation may proceed faster with the free acid (general acid catalysis) than with the ester, we attempted to reverse the last two

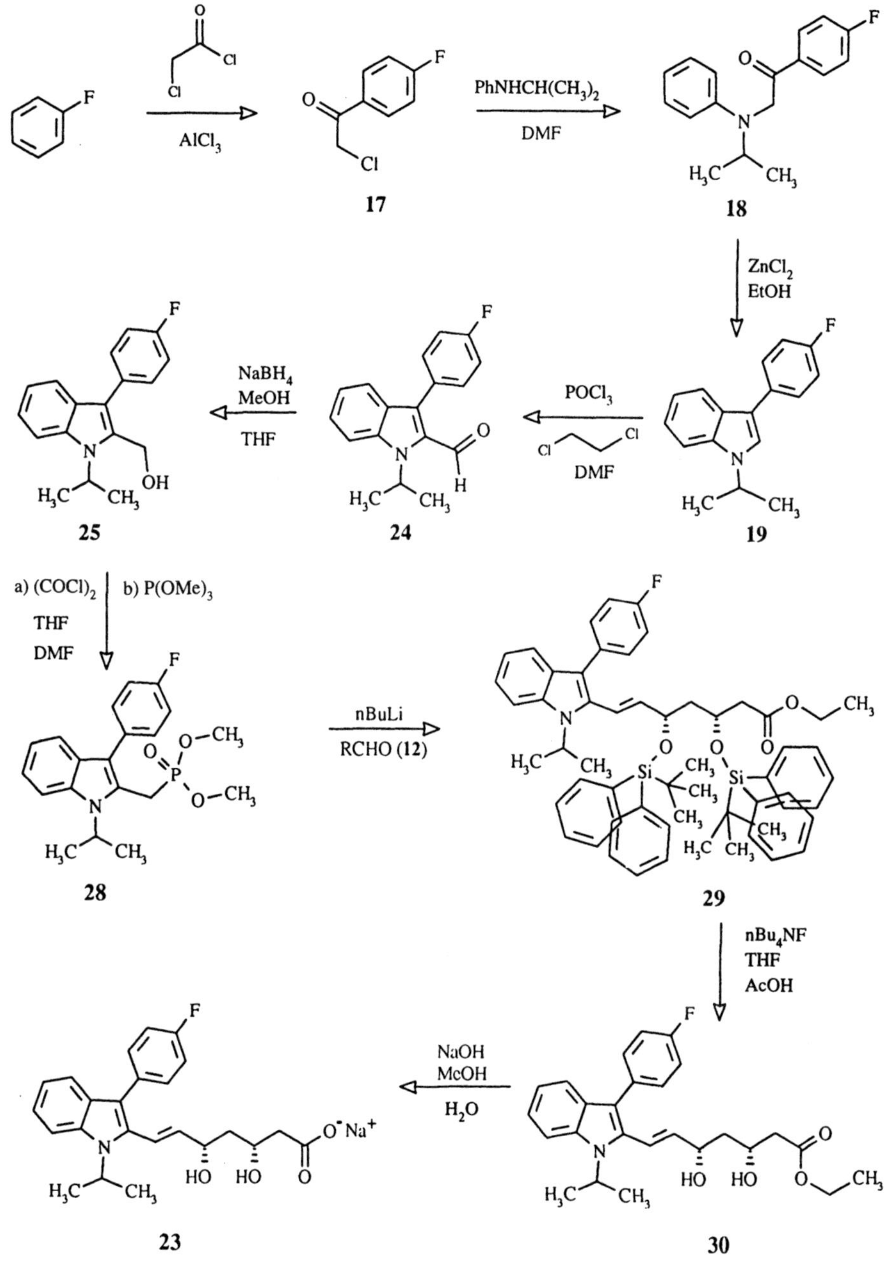

Fig. 3.16. Second, convergent process to fluvastatin.

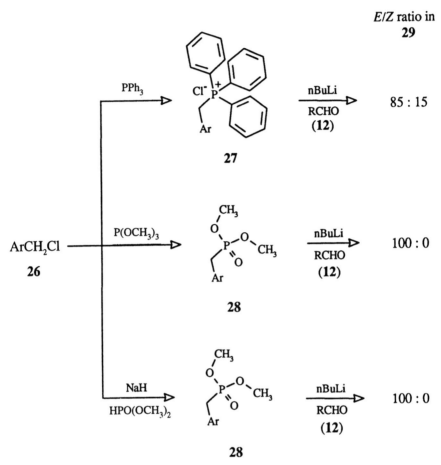

E/Z ratio in **29**

85 : 15

100 : 0

100 : 0

Fig. 3.17. Optimization of stereoselectivity and ecology.

steps of the synthesis, i.e., to perform the saponification before the desilylation. Unfortunately, the silylated diol ethyl ester **29** did not react with sodium hydroxide, and it decomposed under more vigorous conditions (Fig. 3.18).

Consequently, we prepared the *allyl* ester **31** that can be removed under neutral conditions (Fig. 3.19). Its synthesis, of course, involved going back to Prasal: the seven-membered lactone **10** was opened with allyl alcohol instead of ethanol, and the allyl analog of **12** was coupled, as before, in the Horner–Emmons reaction. After a few trials, the removal of the allyl ester was successful (Table 3.7).

The desilylation of *acid* **32**, using six equivalents of tetra-*n*-butylammonium fluoride and acetic acid in THF, indeed proceeded much faster (2 days) than the desilylation of the *ester* **29** (5 days) to give the corresponding dihydroxy acid in excellent yield (90%). Furthermore, no lactone formation was observed under these conditions (11).

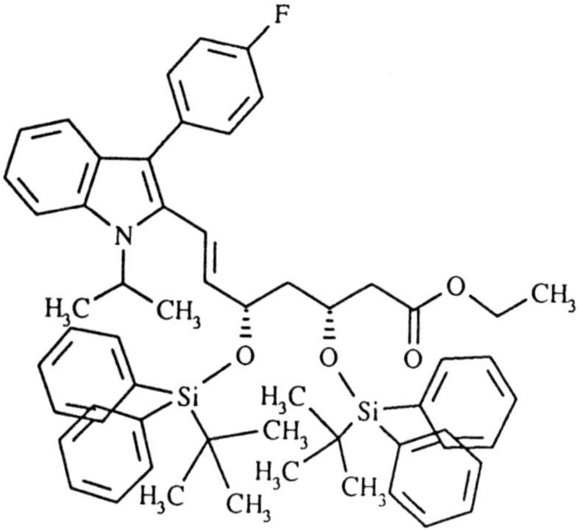

Fig. 3.18. Unsuccessful saponification of ethyl ester.

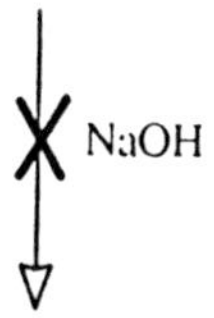

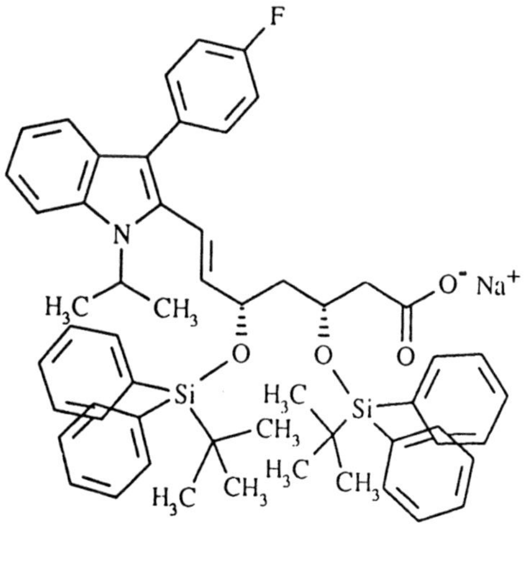

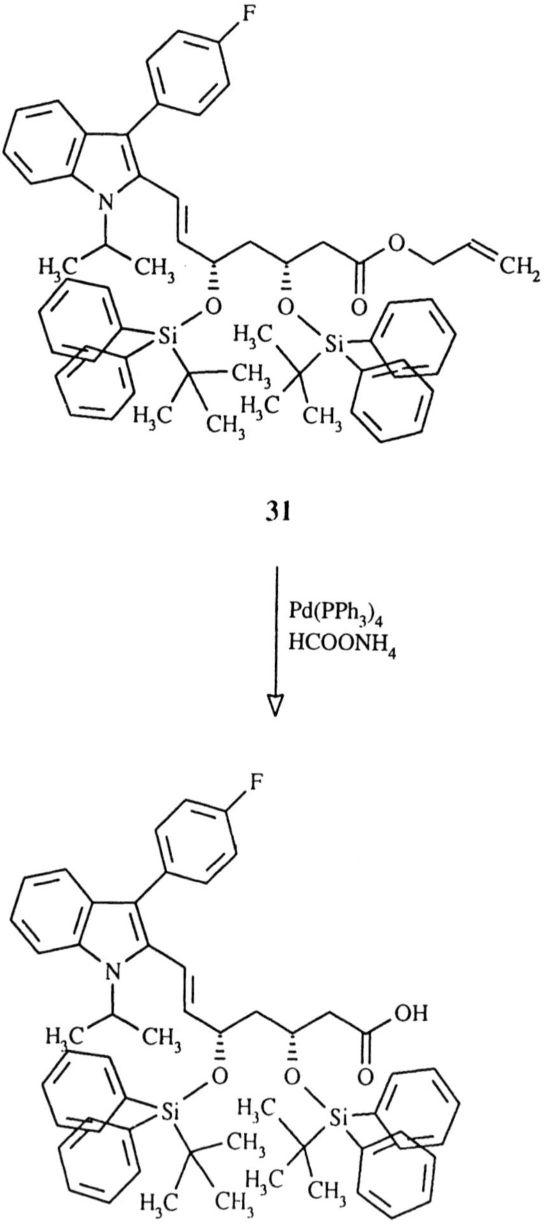

31

Pd(PPh$_3$)$_4$
HCOONH$_4$

32

Fig. 3.19. Removal of allyl ester.

Table 3.7. Removal of the Allyl Group

Reaction Conditions	Yield, %
Pd(Pϕ_3)$_4$, CH$_3$COOH, CH$_2$Cl$_2$, 24°C	50
Pd(OAc)$_2$, Pϕ_3, HCOONH$_4$, dioxane, reflux	10
Pd(Pϕ_3)$_4$, HCOONH$_4$, dioxane, reflux, 1 h	85
Polymer-supported Pd(Pϕ_3)$_4$, HCOONH$_4$, THF, reflux, 18 h	85

3.2.8. Other Protecting Groups

Although there were good reasons for using the bulky *t*-butyldiphenylsilyl protecting group (see Section 3.1.3), it made the synthesis quite expensive. The reagent was large, making the cost per mole high, and we used—and discarded—two equivalents. Therefore, we looked for other protecting groups; the acetonide could be introduced successfully (Fig. 3.20). However, the Horner–Emmons reaction gave only a low yield of the desired olefin, supporting the hypothesis that the α-hydroxyl must be protected by a bulky and hydrophobic group to prevent *hydration* of the aldehyde.

The diphenylsiloxane protecting group was also tried (Fig. 3.21). However, attempts to debenzylate under neutral conditions using hydrogen on 10% Pd/C were unsuccessful. The use of catalytic amounts of acetic acid or HCl resulted in the removal of the diphenylsiloxane protecting group, which could, therefore, not be used in our synthesis.

Although the stereoselective convergent synthesis (process 2) was now well worked out, it was never used in the plant because the synthesis was too long and too expensive. These findings, however, were successfully used to make many analogs of this drug substance. Consequently, we worked on the linear synthesis in

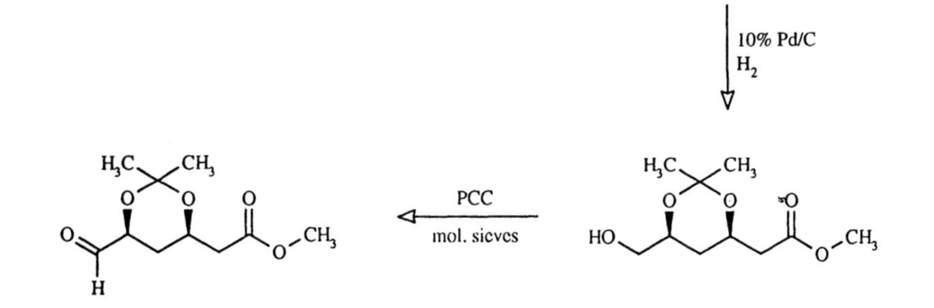

Fig. 3.20. Diol protection by acetonide.

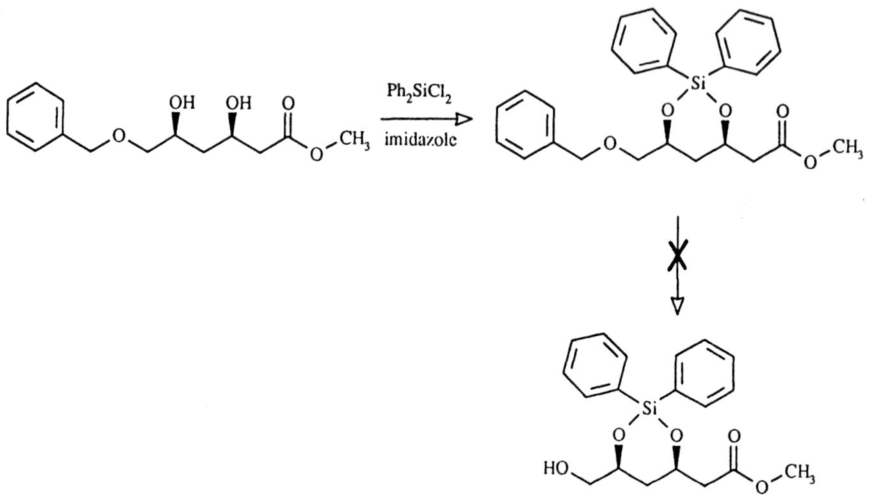

Fig. 3.21. Diol protection by diphenylsiloxane.

parallel and optimized all its steps, persevering especially on the stereoselectivity of the reduction step.

3.3. PROCESS OPTIMIZATION

3.3.1. Intermediate 19

The amination of **17** leading to **18** (Fig. 3.10) was optimized (best results are in parentheses) relative to the temperature (>105°C), time (5 h), and solvent (ethanol). The cyclization step from **18** to **19** was optimized relative to the solvent (ethanol), Lewis acid (zinc chloride), temperature (100°C), and time (4 h). The product **19** could be precipitated with ecologically friendly water, and the reaction yield and purity were both 99%. The cyclization could also be performed in toluene at 150°C in 4 h, using only 1% of ZnCl$_2$ (ecological advantage); however, since the yield was lower (89%) and the required temperature higher (150°C), the ethanol process was preferred. Ethanol was chosen also for the second important reason: Both the amination with N-isopropylaniline leading to **18** and the cyclization to **19** worked well in ethanol, so the two steps were combined, for a total yield of 86%. Some advantages of the new process are apparent from Table 3.8.

The excess N-isopropylaniline (used as a base and as the starting material) was recovered from the aqueous layer (NaOH → pH 11) by toluene extraction in 94% yield and contained 1% aniline, 3% diisopropylaniline, and 6% toluene. It was recycled successfully.

Table 3.8. Comparison of Two Processes to Produce 19

Item	Process 3	Process 1
Number of steps to **19**	2	3
Overall yield	85%	70%
Extraction solvent	Toluene	Dichloromethane
Number of crystallizations	1	2
Purity	99%	98%

3.3.2. Reagent 36

N,N-Dimethyl-3-aminoacrolein was one reagent in the synthesis of **23** (Fig. 3.10) that contributed 67% of the cost of the drug substance. Furthermore, it was not readily available in large quantities. For these reasons, we looked for alternative vinylogous-formylation reagents. The reagents in Fig. 3.22 were tried with varying degrees of success.

The combination of *N,N*-dimethylformamide, oxalyl chloride, and ethyl vinyl ether was used to develop an in-house synthesis of *N,N*-dimethyl-3-aminoacrolein. The three reagents were mixed then quenched with aqueous potassium carbonate and dimethylamine (the latter regenerates N,N-dimethyl-3-aminoacrolein from

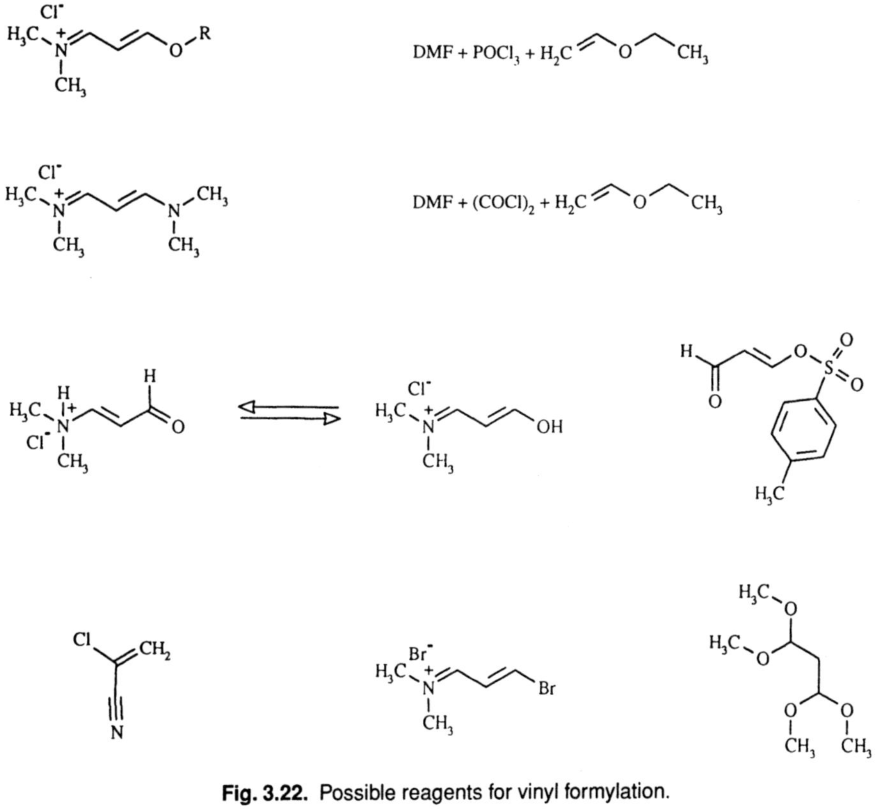

Fig. 3.22. Possible reagents for vinyl formylation.

3-ethoxy-acrolein, which is a major by-product of the hydrolysis) to obtain *N,N*-di-methyl-3-aminoacrolein in a yield of 54% and purity of 88%, which could be used without purification in the preparation of **33**. In this way, the cost was reduced by at least 80%. Among the problems with the use of *N,N*-dimethyl-3-aminoacrolein was its instability under the conditions of the Vilsmeier reaction (leading to **33**): It tended to polymerize and to reformylate the product on the side chain. To obtain the vinylogous aldehyde with a purity of greater than 97%, column chromatography of the crude product was required, but chromatography was not popular for large-scale plant preparations.

An improvement was found (Fig. 3.23) in the reagent *N*-methyl-*N*-phenyl-3-aminoacrolein **36**. The advantages of using **36** instead of *N,N*-dimethyl-3-aminoacrolein are

- Easier preparation of reagent.
- The reagent can be crystallized.
- Shorter reaction time in the formylation step.
- Cleaner **33** (no side-chain formylation).
- No chromatography of **33** is needed.
- Higher yield of **33** is obtained.

Intensive investigations were conducted to develop a large-scale process for reagent **36**. A published procedure used phosgene and ethyl vinyl ether as the

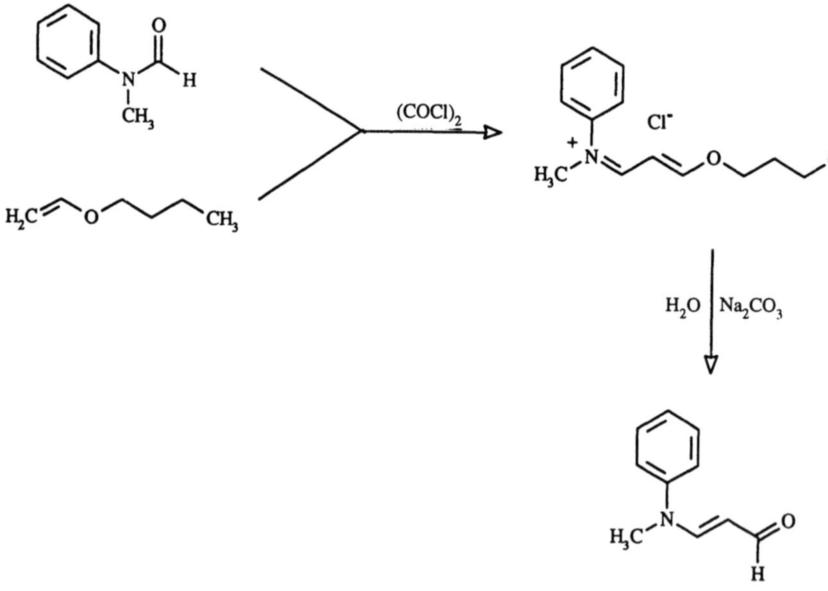

Fig. 3.23. Synthesis of *N*-methyl-*N*-phenyl-3-aminoacrolein.

reagents and dichloromethane as the solvent (12). In our process, phosgene was replaced with oxalyl chloride for safety reasons, ethyl vinyl ether was replaced with butyl vinyl ether for safety reasons, and dichloromethane was replaced with acetonitrile for ecological reasons (13). Many attempts were made to study the scale up, reproducibility, the order of quenching, the order of addition of reagents, temperature, and addition time. Based on these experiments, the order of quench did not show any improvement. However, the order of the addition of reactants was critical for reproducible scale up (yield and purity): A mixture of *N*-methylformanilide and butyl vinyl ether in acetonitrile was added *to* oxalyl chloride at $-5 \rightarrow 10°C$. The process improvements are summarized in Table 3.9.

Another significant discovery was that the crude reagent **36** can be precipitated from isopropanol–hexanes in 82% yield and 99% purity; this procedure avoided the need to form unstable salts. Following this process, we initiated custom manufacture of this reagent.

3.3.3. Intermediate 33

The use of *N*-methyl-*N*-phenyl-3-aminoacrolein (Fig. 3.24) allowed the workup of the vinyl formylation **19** → **33** to be simplified considerably (14). Previously, three filtrations and a silica gel treatment were necessary to separate the polymeric by-products. Now the reaction could be quenched with water to hydrolyze the iminium intermediate to the desired aldehyde and to precipitate the product. The filtrate contained all the by-products (phosphates, methylaniline, and acetonitrile, which were all separated and isolated for disposal or recycling), whereas the filter cake contained only the product. These advantages are summarized in Table 3.10.

The solids were dissolved in toluene, treated with cellulose (to separate insoluble by-products), and crystallized from 95% ethanol (wet ethanol was required otherwise an ethyl acetal formed) to obtain a 75% yield and 99% purity of **33**.

Another significant invention (Fig. 3.25) was to form the aminoacrolein reagent in situ (short of hydrolysis) and subsequently add indole **19** (15). This process

Table 3.9. Comparison of Processes for 3-Aminoacroleins

Item	Process 4	Process 5	Process 6	Process 7
3-Aminoacrolein	N,N-Me$_2$	N-Me-N-ϕ	N-Me-N-ϕ	N-Me-N-ϕ
Volume, mL/g	7	1.5	0.3	0.4
Solvent	CH$_2$Cl$_2$	CH$_2$Cl$_2$	CH$_3$CN	CH$_3$CN
Reagent	(COCl)$_2$	(COCl)$_2$	(COCl)$_2$	COCl$_2$
Vinyl ether	Ethyl	Ethyl	*n*-Butyl	*n*-Butyl
Quench	K$_2$CO$_3$	Na$_2$CO$_3$	Na$_2$CO$_3$	Na$_2$CO$_3$
Anhydrous Na$_2$SO$_4$	Yes	No	No	No
Distillation	Yes	No	No	No
Purity (crude), %	88	88	90	86
Yield (crude), %	57	88	89	83
Cost, $/kg	82	40	57	29

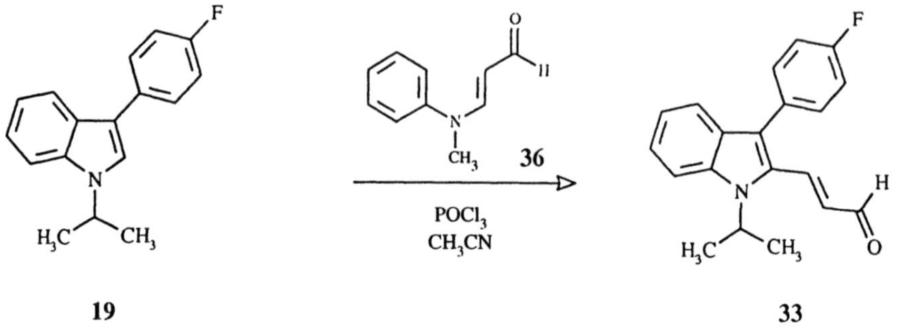

Fig. 3.24. Vinylogous formylation of indole.

eliminated yet another step (or the purchase of the expensive reagent); however, it would require the use of flammable ethyl vinyl ether, so this process was not attractive for the plant.

3.3.4. Intermediate 34

Except for the switch to the *t*-butyl ester (the reasons for which will be explained in Section 3.3.5), no major changes to the reaction conditions (Fig. 3.26) were made at first. One successful attempt to replace THF as a solvent by toluene was carried out, but toluene had its disadvantages in that a slightly lower yield and too slow a formation of the sodium salt of acetoacetate was observed. Replacing THF with DMPU prevented the reaction from occurring, and no product was formed.

Intermediate **34** had not been isolated in the discovery procedure. It was obtained as a crude product dissolved in THF after reaction workup. Since pure starting material was desirable for studying the subsequent reduction of **34** to **35**, attempts to crystallize **34** were initiated. Fortunately, we discovered that intermediate **34** could be recrystallized from toluene–hexane and was much more stable than its crude form: The crystalline powder could be stored at room temperature for months.

Table 3.10. Advantages of *N*-methyl-*N*-phenyl-3-aminoacrolein

Item	N,N-Me$_2$	N-Me-N-ϕ
Ratio **19**/reagent/POCl$_3$	1/2.7/3	1/2.1/2.5
Reaction time at 83°C, h	16	1.5
Side-chain formylation	Yes	No
Silica gel treatment	Yes	No
Purity, %	100	99
Yield, %	67	75

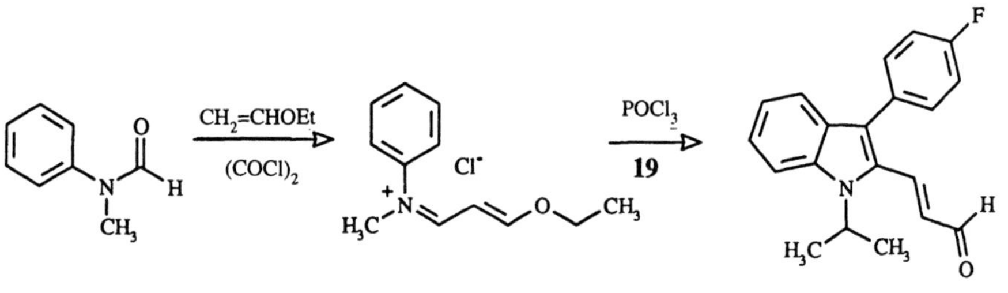

Fig. 3.25. In situ formation of aminoacrolein reagent.

The advantages of isolating and crystallizing **34** were the following:

· The crystalline intermediate was more stable than its solution.
· A pure, defined starting material for reduction step was prepared.
· The purer **34**, the less reagent was needed in the next step.
· The intermediate had a longer storage life.
· Compound **34** could be purified by recrystallization, if necessary.

3.3.5. Intermediate 35

Optimization of the Stereoselectivity. As already mentioned, the isolated yield of the reduction step (Fig. 3.27) on a large scale was only 34%; the goal was to optimize the stereoselectivity of the reduction to such an extent (>98%) that no elaborate purification measures would be needed to separate the diastereomers. It was clear based on literature precedent that a chelating agent would be needed for high stereoselectivity (to fix the acyclic hydroxy–ketone **34** in a ring, which differentiates the two faces of the ketone and allows the reduction to occur from the less hindered face). Consequently, we continued to vary the reducing agent, the chelating agent, the solvent, and the temperature of the reaction and observed the changes in the *syn : anti* ratio in the product (Table 3.11).

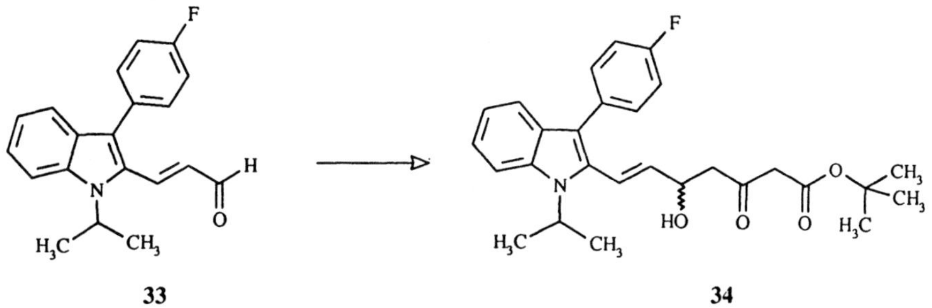

Fig. 3.26. Extension of side chain by four carbon atoms.

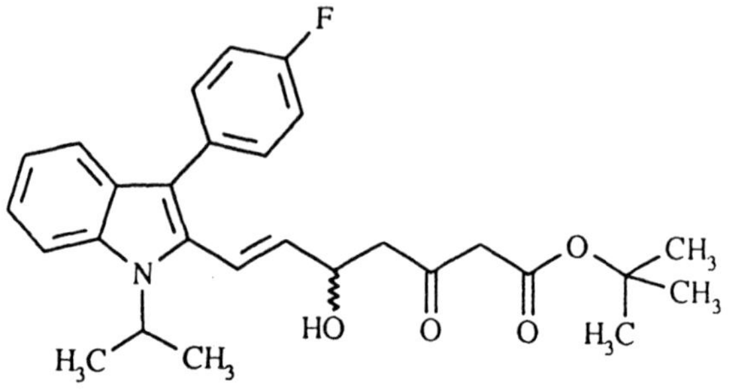

34

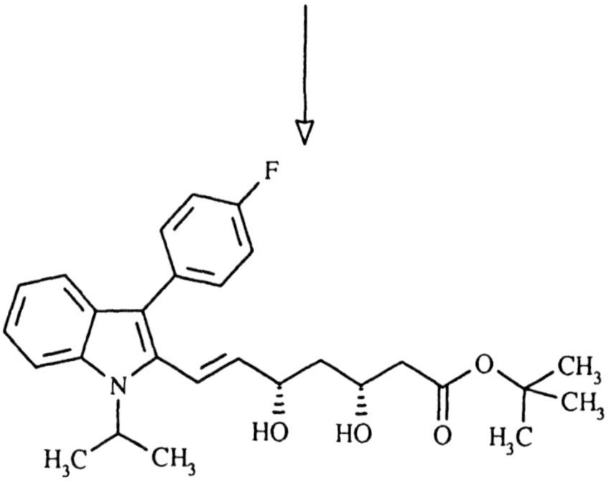

35

Fig. 3.27. Stereoselective reduction leading to *syn*-1,3-diol.

Table 3.11. Optimization of Stereoselectivity

Reducing Agent	Chelating Agent	Solvent	Temperature, °C	*syn*	*anti*
LiBH$_4$	Mg(O$_2$CCF$_3$)$_2$	Et$_2$O	−78	68	32
H$_2$	Pt/C	EtOH	+20	64	36
Me$_2$NH·BH$_3$	Mg(O$_2$CCF$_3$)$_2$	Et$_2$O	−78	41	59
Zn(BH$_4$)$_2$	Zn(BH$_4$)$_2$	Et$_2$O	−20	76	24
NaBH$_4$	Et$_3$B/air	THF	−78	80	20
LiBH$_4$	L-Malic acid	Et$_2$O	−78	63	37
LiBH$_4$	CeCl$_3$	Et$_2$O	−78	55	45
NaBH$_3$CN	—	MeOH	−78	40	60
L-Selectride	*sec*Bu$_3$B + HCO$_2$H	THF	−78	84	16
NaBH$_4$	Et$_3$B + MeOH	THF	−78	98	2

Table 3.12. Discovery of Stereoselective Chelating Agent

Chelating Agent	*syn*	*anti*
Et$_3$B	98	2
Et$_2$BOMe	99	1
EtB(OMe)$_2$	50	50
B(OMe)$_3$	35	65

The results were not encouraging at first as no combination of reagents and chelating agents gave a selectivity that was higher than 80:20, which had already been achieved with NaBH$_4$/Et$_3$B/air in the original discovery synthesis. The breakthrough came when an acid was added to the reaction (e.g., formic acid + *sec*Bu$_3$B); the stereoselectivity rose above 80% for the first time. And when Et$_3$B/air was replaced with Et$_3$B/methanol (also an acid), the selectivity jumped to 98%. Such selectivity was a surprising result, as one would assume a priori that methanol would disrupt the chelation needed for stereoselectivity; instead it improved it dramatically. Clearly methanol was reacting with the Et$_3$B; to define the actual reagent, all combinations of triethylboron and methanol were prepared. One-, two-, and three-ethyl groups were formally replaced with methanol; and these reagents were subjected to the reaction conditions (Table 3.12).

Lo and behold, the best reagent was diethylmethoxyboron (16); it gave a near complete stereoselectivity of 99%. The amount of methanol needed for best results was then varied (Table 3.13), leading to the conclusion that a ratio of THF:MeOH of 4:1 was best.

During the development of this method (17), the diethylmethoxyboron was prepared following Köster's procedure (18). Later, a new method was invented for the in situ generation of Et$_2$BOCH$_3$ without the need for the additional activators (air or acid) (19). However, these conditions were found not to be completely reproducible. Later, commercial suppliers were identified. Initially, the commercial reagent was only 80% pure, but it is now available with higher purity from Aldrich and Callery.

The amount of the chelating agent was also optimized (Table 3.14). The results showed that the more chelating reagent used and the purer the starting material, the higher the stereoselectivity. Thus if both the starting material **34** and the chelating agent were pure, as little as 0.5 equiv of diethylmethoxyboron could be used to still

Table 3.13. Optimization of Solvent Ratio

Solvent Ratio (THF:MeOH)	Isomer Ratio (*syn:anti*)
20:1	94:6
10:1	96:4
6:1	97.5:2.5
4:1	98:2
3:1	97.6:2.4

Table 3.14. Optimization of the Amount of Chelating Agent

Purity of **34**, %	Et$_2$BOMe, equiv	Ratio of *syn : anti* in **35**
67	0.3	92.5 : 7.5
67	0.5	95.8 : 4.2
84	0.5	97.7 : 2.3
67	0.7	96.9 : 3.1
79	0.7	97.9 : 2.1
67	0.9	98.4 : 1.6
84	0.9	99.2 : 0.8
99	0.5	99.1 : 0.9

obtain acceptable stereoselectivity. These studies also explained the mechanism of the reaction: methanol substitutes one ethyl group in Et$_3$B, forming Et$_2$BOCH$_3$. Air in the old, published procedure served a similar function to oxidize one alkyl group, but it was less controllable and thus less reproducible. The hydroxyl group of **34** then reacts with the chelating reagent by displacing the methoxy group on boron; this reaction is more efficient and quicker than with Et$_3$B. The boron is thus bound covalently to **34** and is in place to chelate the ketone intramolecularly. This chelation not only forms a ring that differentiates the two faces of the ketone (resulting in high stereoselectivity) but also activates the ketone for reduction. The chelation is reversible (thus the reason for methanol as the co-solvent) and allows the use of <1 equivalent of the chelating reagent. Apparently, the unchelated ketone does not reduce competitively at the low temperature, otherwise the stereoselectivity would drop when using a deficit of the chelating agent. Other solvent systems did not afford as high a stereoselectivity as tetrahydrofuran (Table 3.15).

Although the stereochemical purity was excellent, the product had to be purified to remove by-products and small amounts of starting material. Since the methyl ester **20** could not be purified by crystallization, the borate derivative **22** (Fig. 3.28) was at first prepared and recrystallized as already described. With the *t*-butyl ester **35**, direct crystallization was possible without derivatization.

Removal of Boron with Hydrogen Peroxide. The new reagents, Et$_3$B or Et$_2$BOCH$_3$, introduced new boron-containing impurities (e.g., the ethylboronate **37**). This by-product may explain why the borate-removing methods (azeotropic distillation with methanol and even the ion-exchange resin) were suddenly not as

Table 3.15. Optimization of Solvent

Solvents (4 : 1)	Ratio (*syn : anti*)
THF : MeOH	98.4 : 1.6
EtOAc : MeOH	92.7 : 7.3
t-BuOMe : MeOH	91.7 : 8.3
ϕMe : MeOH	90 : 10

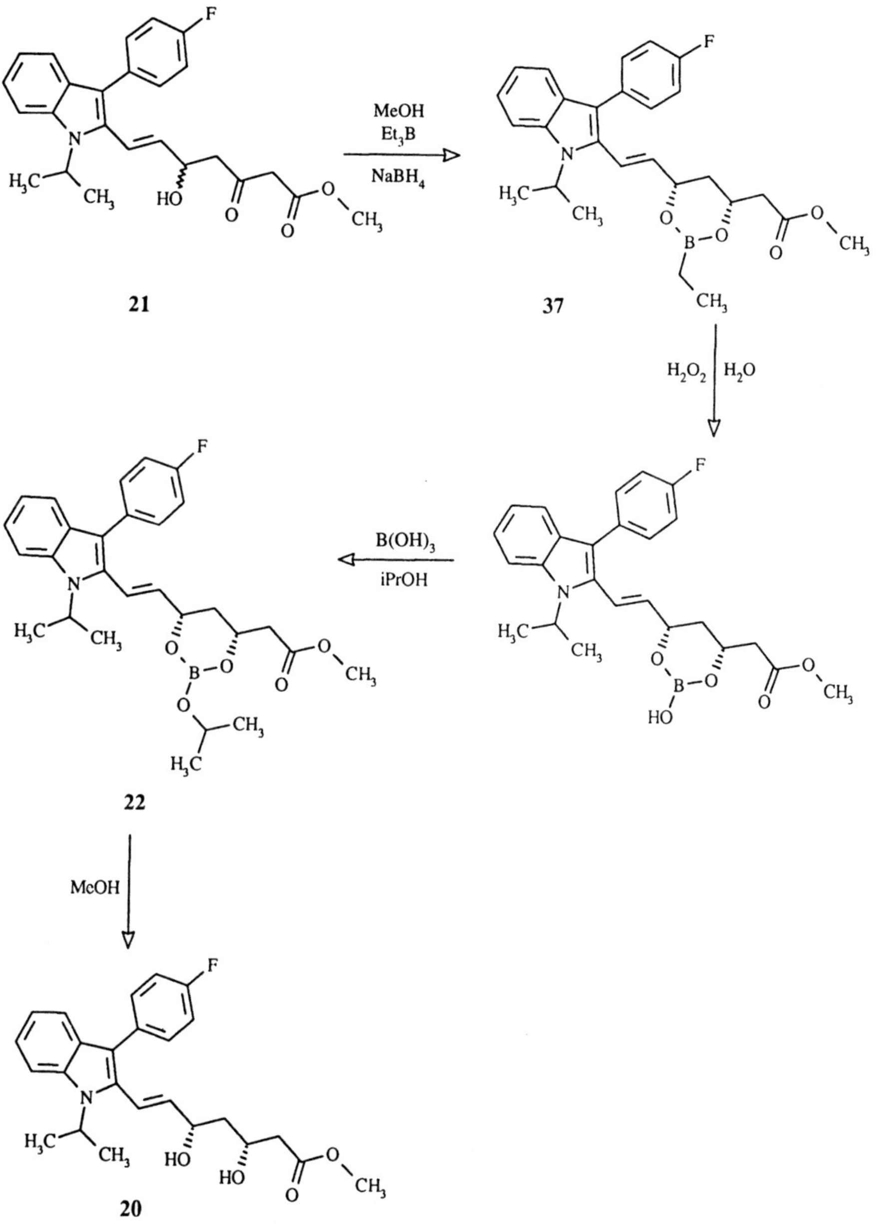

Fig. 3.28. Oxidation of boronate for easier removal of boron.

effective as previously. Consequently, a new operation was added to the workup, namely a reaction of the crude product with aqueous hydrogen peroxide (Fig. 3.28). The peroxide converted any boronate **37** into a borate, which was more easily removed as boric acid (by extraction) or as methyl borate (by distillation with methanol). This improvement allowed us again to reach boron levels of <10 ppm.

Switch to t-Butyl Ester. The reduction reaction **21** → **20** was quenched variously with ammonium chloride/HCl, with acetic acid, or with saturated ammonium chloride. The reason for this trend toward higher and higher pH values was our observation that the diol methyl ester **20** was sensitive to pH in that at low pH it tended to lactonize during solvent evaporation. Even mildly acidic pH was problematic (Fig. 3.29). The lactone furthermore isomerized in the allylic position, which was noticed as higher *anti* content in the drug substance. Although this isomerization could be prevented by keeping the pH of the washes above 7, to be even safer, we replaced the methyl ester with the less reactive *t*-butyl ester in the two steps of the synthesis.

The change to the *t*-butyl ester suddenly presented many advantages:

· No lactone formation and thus no isomerization to the *anti*-isomer.
· Higher yield of the reduction step **34** → **35** (73%).

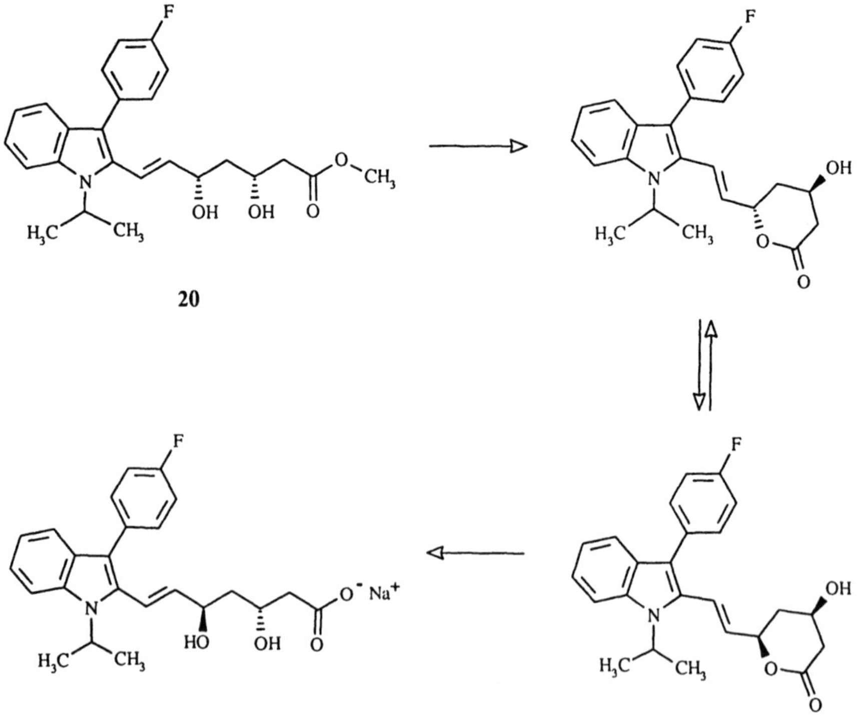

Fig. 3.29. Isomerization via lactonization.

- Higher stereoselectivity (99%).
- Crystalline diol ester **35** and no need for borate derivatization.
- Crystallization of **35** removed boron impurities.
- Purer product **35** (colorless).

The synthesis of **23** was thus shortened to six synthetic steps with no chromatography (Fig. 3.30). Considering the complexity of the structure, this process was quite an accomplishment (20).

Sequence of Reagents. Normally and previously, the reagent sodium borohydride was added to the reaction mixture last. However, knowing the sensitivity of Et$_2$BOMe to air and peroxides and accepting the need to use commercial solvents without purification, a decision was made to alter the sequence of addition of reagent and reactant. That is, sodium borohydride was added to the solvents (THF/MeOH) first; it presumably cleaned the solvents of peroxides, aldehydes, and other interfering impurities (and it may even activate NaBH$_4$ as it slowly reacts with methanol). Ketone **34** was added last. This sequence brought the following advantages: instantaneous reduction of the substrate, further enhancement of the *syn*-stereoselectivity to >99%, and greater reproducibility of the results.

Since the pilot plant was concerned about adding sodium borohydride to methanol (in case of cooling failure, the hydrogen evolution would become uncontrollable), they added the methanol with the ketone, but the stereochemical results were nevertheless still acceptable.

Work-up Procedure. As already mentioned, the advantage of the *t*-butyl ester was that diol **35** was crystalline, and the derivatization with boric acid was not needed anymore. Several solvents were tested to crystallize **35** (Table 3.16).

Ethyl acetate was chosen as a compromise between yield and toxicity (both of which are higher with acetonitrile). At first, two crystallizations from ethyl acetate were needed to meet all specifications, and two recrystallizations of the combined second crops (Table 3.17).

Compared to the old process (five crystallizations of the borate), this purification was an improvement, but it still involved too many operations. With further optimization, **35** was isolated efficiently in one crop, involving the peroxide treatment, an ethyl acetate–heptane slurry, and one crystallization from ethyl acetate. The total yield for the two synthetic steps was 73%.

Thus a systematic approach to a reproducible process (Table 3.18) successfully lowered the boron content to 7 ppm and increased the purity from 80.7 to 98.4%.

In summary, the latest process incorporated the following improvements: the molar ratio of **34**, Et$_2$BOMe, and NaBH$_4$ was optimized at 1:0.7:1.22; ethyl acetate replaced heptane as the extraction solvent (to minimize the number of different solvents in the process); intermediate **35** was crystallized in one crop from ethyl acetate after an ethyl acetate–heptane slurry; and sodium chloride solution in the extractions was left at its natural pH of 5–6, as the lactone formation was no longer a problem with the *t*-butyl ester.

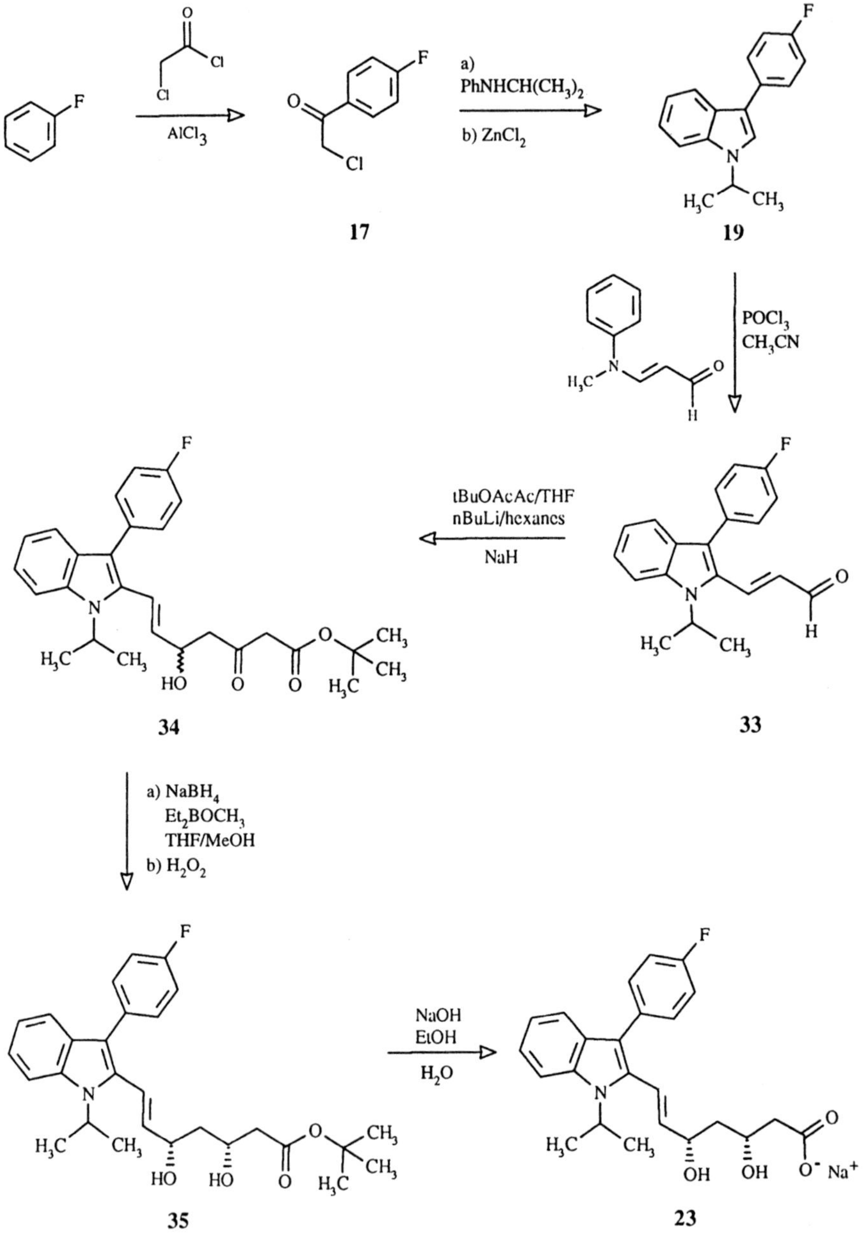

Fig. 3.30. Final synthesis of fluvastatin.

Table 3.16. Optimization of Crystallization Solvent

Solvent	Temperature, °C	Recovery, %
Ethyl acetate	0	89
Acetonitrile	20	94
Toluene	20	82
Isopropanol	20	80

It may be apparent by now that process optimization is an open-ended activity. The final process added the following improvements: Crystalline **34** was used, therefore the molar ratio of **34** and Et$_2$BOMe could be reduced to 1:0.5; sodium peroxide was used to oxidize the boronate, and the amount of peroxide could thus be reduced by 70% (sodium perborate or percarbonate was also used successfully in the laboratory); the amount of THF/MeOH was reduced by 20%; the amount of sodium bicarbonate was reduced by 50%; the amount of ethyl acetate was reduced by 15%; the amount of sodium chloride solution was reduced by 66%; and the amount of sodium sulfite was reduced by 45%. The yield was now 73%, and the purity of this drug substance precursor (**35**) was excellent.

3.3.6. Drug Substance 23

Solvent for Saponification. The drug substance was at last prepared (see Fig. 3.30) by hydrolyzing the ester at room temperature with aqueous sodium hydroxide (exactly 1 equiv, as any excess would end in the drug substance). The methyl ester was at first hydrolyzed in methanol as the co-solvent; since a less toxic solvent was preferred in the last step of the synthesis, methanol was successfully replaced with ethanol. When the synthesis switched to the *t*-butyl ester, we noticed that **35** was not soluble in alcohols and the saponification occurred too slowly. Consequently, tetrahydrofuran was used as a co-solvent for the saponification. The aqueous layer was then lyophilized to give the pure drug substance. Because there was little purification performed in this step (only extraction with an organic solvent), we had taken much effort to purify the precursor **35**.

The hydrolysis of the *t*-butyl ester was surprisingly fast; we believe the hydroxyl groups activate the hydrolysis (intramolecular general acid catalysis). This notion is supported by the fact that by removing or protecting first the δ-, then the β-hydroxyl group, the saponification becomes progressively more difficult (see Fig. 3.18).

Table 3.17. Purification by Recrystallization from Ethyl Acetate

Quality Specification	First EtOAc	Second EtOAc
Syn : anti ratio	99.28:0.72	99.73:0.27
Precursor, %	0.2	0
Boron, ppm	123	20
Appearance	Off-white	White
Yield (first crop), %	68	54

Table 3.18. Purification Sequence

Operation	Purity, %	Ratio (*syn* : *anti*)	Boron, ppm
H_2O_2 treatment	80.7	97.89 : 2.11	51.5
EtOAc–heptane slurry	87.5	98.06 : 1.94	18.8
EtOAc crystallization	98.4	99.35 : 0.65	7.1

Optical Resolution. Lescol contains a racemic drug substance. Daicel Chemical Industries, Ltd. (Japan) developed for us a chromatographic separation of the enantiomers of **20** on their optically active column Chiralcel OF. The optical rotations for **20** were $[\alpha]_D + 54$ ($c = 0.5$, methanol) and $[\alpha]_D - 53$ ($c = 0.5$, methanol) and for **23**, $[\alpha]_D + 27.5$ ($c = 0.8$, methanol) and $[\alpha]_D - 26.3$ ($c = 0.8$, methanol).

The resolution was convenient for toxicological and pharmacokinetic studies; however, it would not be feasible or economical on a commercial scale. In the next three chapters, we will see how to prepare enantiopure drug substances.

BIBLIOGRAPHY

1. U.S. Pat. 5,354,772 (Oct. 11, 1994), "Indole Analogs of Mevalonolactone and Derivatives Thereof," F. Kathawala (to Sandoz).

2. U.S. Pat. 4,739,073 (Apr. 19, 1988), "1,3-Disubstituted-2-carboxy-dihydroxyalkylindole Derivatives, Useful as Hypolipoproteinemic and Atherosclerotic Agents," F. G. Kathawala (to Sandoz).

3. F. G. Kathawala, "HMG-CoA Reductase Inhibitors: An Exciting Development in the Treatment of Hyperlipoproteinemia," *Med. Res. Rev.*, **11**, 121–146 (1991).

4. R. E. Walkup and J. Linder, "2-Formylation of 3-Arylindoles," *Tetrahedron Lett.*, **26**, 2155–2158 (1985).

5. K. Prasad and O. Repič, "A Novel Diastereoselective Synthesis of the Lactone Moiety of Compactin," *Tetrahedron Lett.*, **25**, 2435–2438, (1984).

6. U.S. Pat. 4,841,071 (June 20, 1989), "New 7-Member Ring Lactone Compounds—Useful as Intermediates for Antihypercholesterolemic Agents," K. Prasad (to Sandoz).

7. U.S. Pat. 4,571,428 (Feb. 18, 1986), "6-Substituted-4-hydroxytetrahydropyran-2-ones," K. Prasad (to Sandoz).

8. F. G. Kathawala, B. Prager, K. Prasad, O. Repič, M. J. Shapiro, R. S. Stabler, and L. Widler, "Stereoselective Reduction of δ-Hydroxy-β-ketoesters," *Helv. Chim. Acta*, **69**, 803–805 (1986).

9. J. Dale, "The Reduction of Symmetrical 1,2- and 1,3-Diketones with Sodium Borohydride, and the Separation of Diastereoisomeric 1,2- and 1,3-Diols by Means of Borate Complexes," *J. Chem. Soc.*, **1961**, 910–922.

10. L. H. Horsley, "Azeotropic Data III," *Adv. in Chem.*, **116** (1973).

11. G. T. Lee, J. Linder, K.-M. Chen, K. Prasad, O. Repič, and G. E. Hardtmann, "A General Method for the Synthesis of *syn*-(E)-3,5-Dihydroxy-6-heptenoates," *SynLett.*, **1990**, 508.

12. Brit. Pat. 945,536 (1968), "Preparation of β-Aminoacroleins" (to Instituto Chemioterapico Italiano).

13. U.S. Pat. 5,290,946 (Mar. 1, 1994), "Processes for the Synthesis of 3-(Substituted Indol-2-yl)propenaldehydes," G. T. Lee, P. K. Kapa, and O. Repič (to Sandoz).

14. G. T. Lee, J. C. Amedio Jr., R. Underwood, K. Prasad, and O. Repič, "Vinylformylation Utilizing Propeniminium Salts," *J. Org. Chem.*, **57**, 3250–3252 (1992).

15. U.S. Pat. 5,118,853 (June 2, 1992), "Processes for the Synthesis of 3-Disubstituted Aminoacroleins," G. T. Lee and O. Repič (to Sandoz).

16. U.S. Pat. 5,189,164 (Feb. 23, 1993), "Process for the Synthesis of *syn*-(E)-3,5-Dihydroxy-7-substituted Hept-6-enoic and Heptanoic Acids and Derivatives and Intermediates Thereof," K. Prasad and K.-M. Chen (to Sandoz).

17. K.-M. Chen, G. E. Hardtmann, K. Prasad, O. Repič, and M. J. Shapiro, "1,3-*syn*-Diastereoselective Reduction of β-Hydroxy Ketones Utilizing Alkoxydialkylboranes," *Tetrahedron Lett.*, **28**, 155–158 (1987).

18. R. Köster, W. Fenzl, and G. Seidel, "Katalysierte Dialkylborylierung von Alkoholen und Phenolen," *Liebigs Ann. Chem.*, **1975**, 352–372.

19. K.-M. Chen, K. G. Gunderson, G. E. Hardtmann, K. Prasad, O. Repič, and M. J. Shapiro, "A Novel Method for the In Situ Generation of Alkoxy-Dialkylboranes and Their Use in the Selective Preparation of 1,3-*syn*-Diols," *Chem. Lett.*, **1987**, 1923–1926.

20. Eur. Pat. 363 934 (1990), "Process for the Preparation of 7-Substituted Hept-6-enoic and Heptanoic Acids and Derivatives and Intermediates Thereof," K.-M. Chen, K. Prasad, G. Lee, O. Repič, P. Hess, and M. Crevoisier (to Sandoz).

4

RESOLUTIONS*

The previous chapter described process development of a racemic compound. In the future, however, drug substances will frequently be enantiopure. The reason for this is that the medicinal chemists are performing rational drug design. In other words, they study the mechanisms of diseases and then design small molecules that interfere with these mechanisms, usually by binding to a biochemical mediator. Since these biochemicals are usually chiral, an enantiomerically pure molecule will often fit and dock better than an achiral molecule (three-point interaction model) (1). Actually, one theory predicts that the bigger the difference in activities between enantiomers, the better the structure–activity has been optimized (2).

The realization that two enantiomers can have different activities or even different indications (see Chapter 5) is then the reason why more drugs will be enantiomerically pure. The active enantiomer is called the eutomer (3), the inactive or undesirable enantiomer is the distomer. Ideally, all the desired activity is in the eutomer and all the undesirable side effects or lack of activity in the distomer; and the two should, therefore, be separated.

Enantiomerically pure molecules represent a challenge to process chemists. Enantioselective syntheses are usually longer and more sensitive (e.g., to racemization; see Chapter 2) than achiral syntheses. They are a challenge also because, in spite of tremendous advances in asymmetric synthesis in the past 10 years, there are still few reactions that produce enantiomeric excesses (ee) of >98%. The high standards of the process chemists and the ideal process require nothing less. The chemist thus must invent better methods or optimize known reactions to obtain >98% ee.

*Based on a lecture I gave at the Pharmaceutical Manufacturers Association meeting, Charlottesville, Va., Apr. 19, 1988.

Fortunately, there are many ways of preparing nonracemic substances. Generally, these ways can be categorized as (*1*) resolution, (2) asymmetric synthesis (in which optical activity is induced by an enantiopure reagent), and (*3*) the chiral pool, using optically active natural products as building blocks (chirons). Chapters 4, 5, and 6, respectively, will illustrate these various strategies.

The last of these methods, buying or isolating the building blocks from nature, is obvious but requires two caveats. Nonracemic natural products are not necessarily optically pure; neither nature nor suppliers are as particular and demanding about enantiomeric purity as process chemists are. The other observation is that the chiral pool strategy is usually less efficient than is an asymmetric synthesis. An asymmetric synthesis can be as short as a racemic synthesis—one reagent is just replaced with a transient chiron—whereas a synthesis starting with a natural product may be very winding as one laboriously tries to invert, transform, cut, or otherwise morph the given starting material and its stereogenic centers into the desired product (see Chapter 6 for such an example).

4.1. CHIRALITY

All organic molecules (and all objects, for that matter) can be divided into *achiral* and *chiral* (Fig. 4.1). The word *chiral* literally means handed, as in left- and right-handed. A chiral compound is recognized by the property that its mirror image is nonsuperimposable with the compound; such pairs of compounds are enantiomeric to each other and are called enantiomers. If the mirror image is superimposable, then the mirror image is the same compound as the original, no enantiomer can exist, and the compound is said to be achiral. There is nothing magic about mirrors; they are simply a convenient and intuitively simple way of checking for symmetry in the molecule. The following definition is scientifically better but harder to apply: If the molecule contains an inverse symmetry element (center, plane, or improper axis of symmetry), it is achiral; otherwise it is chiral.

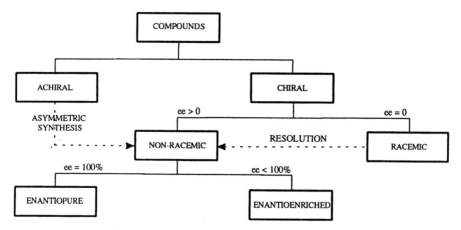

Fig. 4.1. Classification of compounds by chirality.

If a chiral compound contains both enantiomers in the ratio of 1 : 1 (enantiomeric excess = 0), it is *racemic*. If it contains an excess of one enantiomer (enantiomeric excess > 0), it is *nonracemic*. Nonracemic compounds that contain only one enantiomer (enantiomeric excess = 100%) are called *enantiopure* (not *homochiral*, which refers to *different* molecules or crystals of the same chirality). There is no good term for describing compounds that contain a mixture of enantiomers in a ratio other than 1 : 1 (100 > ee > 0). The proposed term *scalemic* has not been accepted or used by the profession, and the term *enantiomerically enriched* is an awkward but preferred descriptor at this time (1). Sometimes otherwise respectable chemists use the term *chiral* when they mean *nonracemic*. This is not correct, as Figure 4.1 shows, because racemic compounds are also chiral. The established terms *chiral auxiliary, chiral pool,* and *chiron* are, therefore, also not correct; but they will be used in this book.

Why do racemates exist as a 1 : 1 mixture of enantiomers? This balance seems odd, so how does it come about? Organic reactions that produce chiral compounds generate both enantiomers in exactly equal amounts. The reason for this is simply that the energy required for the reaction is the same as for its mirror image, so both enantiomers form. Also, two enantiomers have the same physical properties. How can we ever tell whether we have a mixture of enantiomers or a single enantiomer? It turns out that *in a nonsymmetric environment,* enantiomers behave like diastereomers, i.e., they can have different physical and biological properties.

This observation answers the following questions:

1. *Why are pharmaceutical chemists interested in separating enantiomers?* Since biological systems (enzymes) are such asymmetric environments, enantiomers will have differing biological properties. These properties may be synergistic or antagonistic, one enantiomer may be more toxic, the other one more active; whatever the case, we should prepare and know the properties of each enantiomer separately.

2. *How does one separate or distinguish enantiomers?* By placing them in an asymmetric environment, enantiomers can behave like diastereomers, which now have different physical properties and can thus lead to separation by classical means, e.g., by crystallization, distillation, and chromatography. By the way, plane-polarized light is such an asymmetric environment: Enantiomers rotate plane-polarized light in opposite directions; this is how one can distinguish them (D or L, + or −), and this is the origin of the term *optically active*.

As already mentioned, there are two ways of making nonracemic compounds: from racemates by separating enantiomers, and from achiral compounds by asymmetric synthesis.

4.2. METHODS OF RESOLUTION

Methods for separating enantiomers can be divided by the nature of the asymmetric environment: In resolution by *preferential crystallization,* the asymmetric environment is a small excess of one enantiomer that is added before crystallization (seeding); the asymmetry can be on the *chromatography* column; one can form *diastereomeric* derivatives by attaching enantiopure reagents covalently or as salts; or the asymmetry can be in the reagent (*kinetic* resolution). There are two types of resolution that apparently do not require an asymmetric agent; these are appropriately named *spontaneous* resolutions.

4.2.1. Resolution by Preferential Crystallization

By seeding a solution, supersaturated with a racemic compound, with a pure enantiomer, the crystallization of that excess enantiomer can sometimes occur. One can seed with both enantiomers in different locations and obtain separated enantiomers in the same crystallizing dish (Fig. 4.2*a*) (4).

One can obtain both isomers in a single container also by seeding one enantiomer with large crystals and the other enantiomer with small crystals (Fig. 4.2*b*) (5). After crystallization, the separation of enantiomers can be achieved by sifting, as the two will differ in crystal size. Most practically, one can crystallize both enantiomers by placing seeds of each in two separate columns and passing a saturated solution of the racemic substance slowly over both columns in parallel (Fig. 4.2*c*) (6).

The best technique of resolution by preferential crystallization is as follows (7). To a saturated racemic solution is added a small amount of one, let us say (−)-enantiomer. The solution is heated until all solids dissolve, then it is cooled and seeded with the (−)-isomer. Pure (−)-isomer is recovered after crystallization. This exact amount is replenished in the mother liquors with the racemate, making the solution saturated again but now containing an excess of the (+)-isomer. On heating, cooling, seeding with the (+)-isomer, and crystallizing, the (+)-isomer is isolated. This is repeated indefinitely; e.g., with hydrobenzoin, after 15 cycles the original amount of racemate is completely resolved (Table 4.1).

The conditions for such a resolution (concentration, temperature, stirring rate, time of crystallization, and the amount of the initial enantiomeric excess) can be designed and optimized with the help of physical chemistry, e.g., with phase dia-

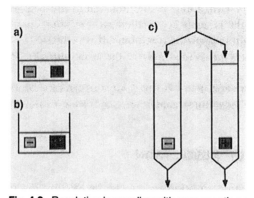

Fig. 4.2. Resolution by seeding with pure enantiomer.

Table 4.1. Procedure for Resolution by Preferential Crystallization[a]

Cycle Number	Added Racemate	Added (−)	Isolated (−)	Isolated (+)
1	11.0	0.37	0.9	na
2	0.9	na[b]	na	0.9
3	0.9	na	0.8	na
4	0.8	na	na	0.75
5	0.75	na	0.7	na
6	0.7	na	na	0.75
7	0.75	na	0.8	na
•	•	•	•	•
•	•	•	•	•
•	•	•	•	•
15	23.5	0.37	6.5	5.7

[a]Reprinted by permission from Ref. 8.
[b]na = not applicable; bullets indicate "etc."

grams, solubility curves, and crystallization rate. The physical chemistry behind this is too complex to go into here, but let me point out one important condition: This method works only for racemates that form *conglomerates,* i.e., they crystallize as a mixture of pure enantiomers, as opposed to a *racemic compound.* The method is thus not general as only about 10–20% of organic compounds form conglomerates; but it is efficient when it works.

4.2.2. Resolution in a Supercooled Melt

As with supersaturated solutions, one can also perform resolutions with seeding of supercooled melts (9). A racemic melt is placed in both tanks 1 and 2 (Fig. 4.3). After supercooling, i.e., a few degrees below the freezing point of the racemate, both are seeded, one with (+), the other with the (−)-isomer.

After crystallization, the liquid is filtered off into tank 11. This mother liquor contains an excess of the (−)-isomer and is pumped into container 2 where the (−)-isomer was crystallized. Similarly with the other mother liquor in tank 12, richer in the (+)-isomer. After several (e.g., seven) cycles, the enantiomers are collected in about 85% purity and a yield of 20% each (the example is for α,β-dimethyl-β-lactam).

4.2.3. Resolutions with Optically Active Solvents

Solvent may be the asymmetric environment for resolving enantiomers. For example, heptaheterohelicene **1** (Fig. 4.4) was resolved by crystallization from (−)-α-pinene (10). Again, the requirement is that the racemate is a conglomerate. The heptaheterohelicene **2**, e.g., is not and cannot be resolved in this manner.

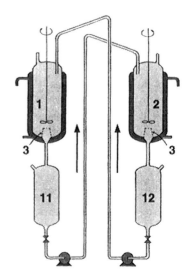

Fig. 4.3. Resolution by seeding a supercooled melt. Reprinted by permission from Ref. 8.

4.2.4. Resolution by Chromatography

One can resolve racemates by chromatography. As the stationary phase, one can use silica gel that is covalently bound to amino acids or impregnated with enantiopure acids (e.g., tartaric acid, malic acid, and camphorsulfonic acid); optically active cyclodextrins; or cellulose functionalized with acetate, carbamate, cinnamate, or toluate groups. The latter was used to resolve mephobarbital **3** (Fig. 4.5) (11); another example can be found at the end of Chapter 3.

4.2.5. Resolution via Diastereomers

The most common and classical resolution is the method of forming diastereoisomeric derivatives: An enantiopure reagent is used to bind to the racemic mixture of enantiomers, either covalently or as a salt. The resulting pair, which is now diastereomeric and no longer related as mirror images, can be separated by ordinary separa-

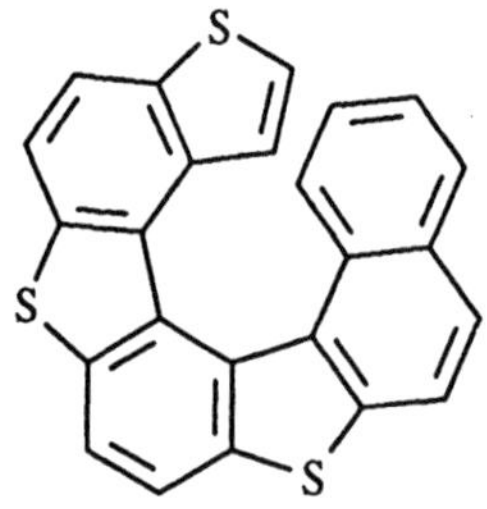

1 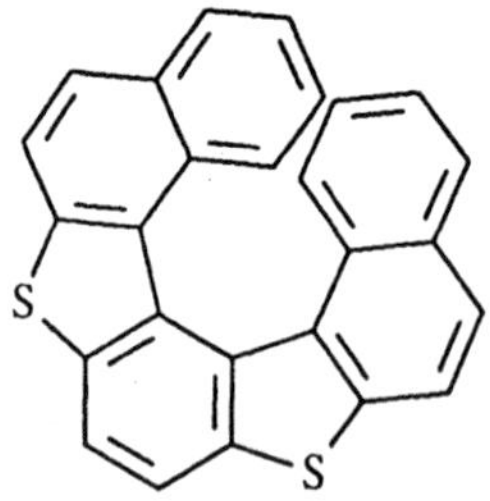   **2**

Fig. 4.4. Structures of heptaheterohelicenes.

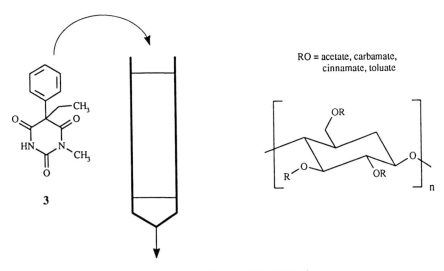

Fig. 4.5. Resolution by chromatography.

tion methods, e.g., by crystallization, distillation, sublimation, and extraction. The enantiopure reagent is then removed, revealing the resolved enantiomer, and recycled. Such a resolution is illustrated with (chiral!) mittens in Fig. 4.6.

For resolving carboxylic acids, one usually forms salts with optically active amines, most often natural products like quinine or other *cinchona* alkaloids. The reason why these are most often used is that they meet most of the requirements of an ideal resolving agent (12):

· Proximity of stereogenic centers on agent and racemic substance;
· Rigid structure;
· Strong acid or base;
· Chemical and optical stability;
· Both enantiomers available; and
· Recyclable.

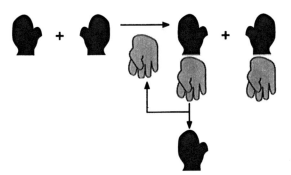

Fig. 4.6. Resolution by forming diastereomeric derivatives.

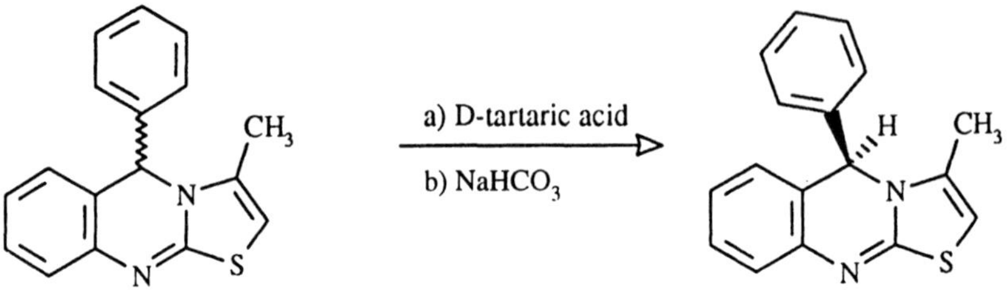

Fig. 4.7. Resolution with D-tartaric acid.

For resolving amines, on the other hand, one uses enantiopure acids like tartaric acid, malic acid, and mandelic acid (Fig. 4.7) (13). To resolve neutral compounds, one prepares covalent, diastereomeric derivatives. For example, with alcohols one can form a camphorsulfonate (14); with ketones one can form hydrazones (15) (Fig. 4.8).

4.2.6. Kinetic Resolution

Another method of resolving enantiomers is kinetic resolution. This method differs from classical resolution in that the enantiopure reagent reacts with not both but only one enantiomer; in other words, it is an enantiospecific reaction (Fig. 4.9).

The selectivity happens because the transition state energy of the reaction of one enantiomer with the optically active reagent will be *different* from the other, the two transition states being diastereomeric. Thus the term *kinetic* resolution.

The most common optically active reagent is an enzyme. A lipase can, e.g., hydrolyze only one enantiomeric ester, converting it to an enantiopure alcohol, which can then be easily separated from the remaining enantiopure ester (Fig. 4.10).

4.2.7. Resolutions with 100% Theoretical Yield

The advantages of classical resolutions are that they have a broad scope, that both reagent enantiomers are usually available (unlike enzymes), and that they can be scaled up. The limitations of chemical resolutions, on the other hand, are that they are unpredictable (trial and error), involve a complex process, and have a low yield.

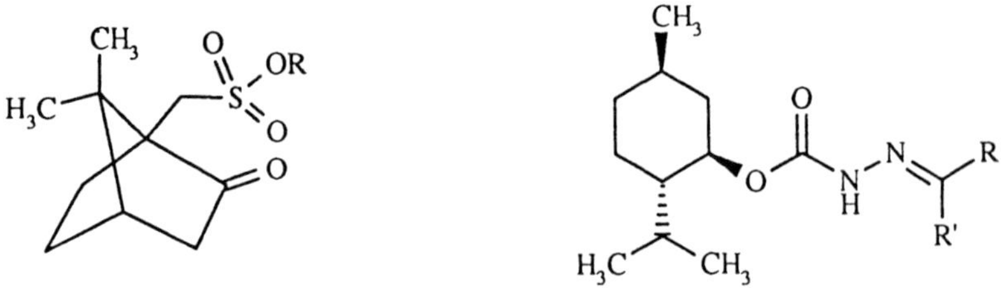

Fig. 4.8. Derivatization of neutral molecules for subsequent resolution.

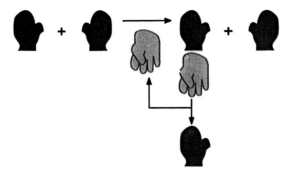

Fig. 4.9. Kinetic resolution with recycling of the enantiopure reagent.

The inherently low yields of resolutions (the theoretical yield is only 50%) can be increased to nearly 100% by various tricks. A few examples follow.

The best resolutions are those in which the undesirable enantiomer (distomer) can be racemized and recycled; yields approaching 100% are then possible. For example, the unnatural D-phenylglycine, needed in the synthesis of ampicillin, is made by a kinetic resolution of DL-phenylglycinamide (Fig. 4.11) (16). The enzyme aminopeptidase hydrolyzes the L-enantiomer and leaves the desired D-phenylglycinamide unchanged. It is separated and chemically hydrolyzed to D-phenylglycine. The undesirable L-phenylglycine is racemized with sulfuric acid and converted to DL-phenylglycinamide with ammonia, and the cycle repeats.

In another example, the undesirable enantiomer was racemized with either light or heat and recycled (Fig. 4.12) (13).

Even more economical are resolutions in which the wrong enantiomer can be racemized and recycled in situ (Fig. 4.13). Hydantoinase hydrolyzes only one enantiomer of hydantoin **4** (17), while the other enantiomer spontaneously racemizes under the reaction conditions, and the entire sample can be converted into the D-isomer. Enantiopure amino acids can thus be produced on a ton scale.

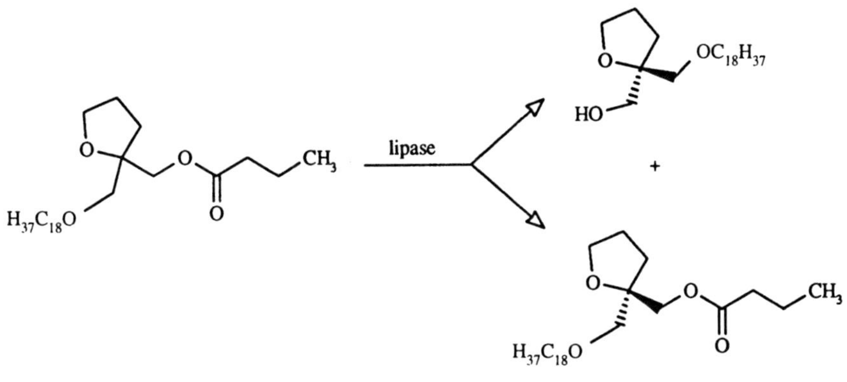

Fig. 4.10. Kinetic resolution with an enzyme.

NH3
DL-phenylglycinamide
DL-phenylglycine
aminopeptidase
H2SO4
D-phenylglycinamide
+
L-phenylglycine
D-phenylglycine
Ampicillin

Fig. 4.11. Racemization and recycling of distomer.

CH3
N
S
200 °C
or
hν

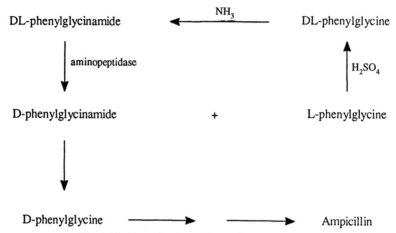

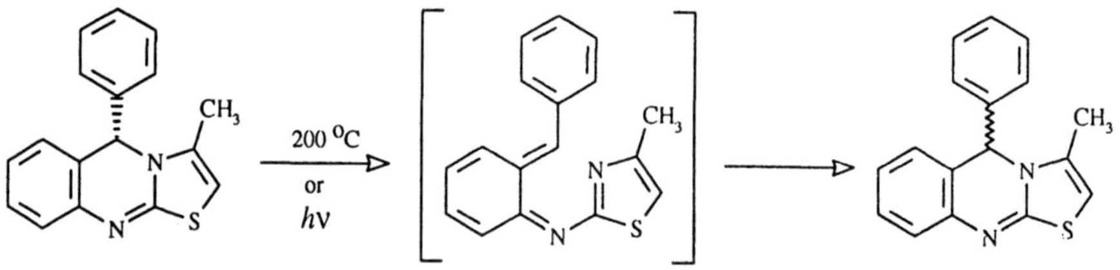

Fig. 4.12. Racemization of distomer by heat or light.

HO
NH
O
N
H
DL - 4
hydantoinase
OH
NH2
D
+
L
spontaneously racemizes

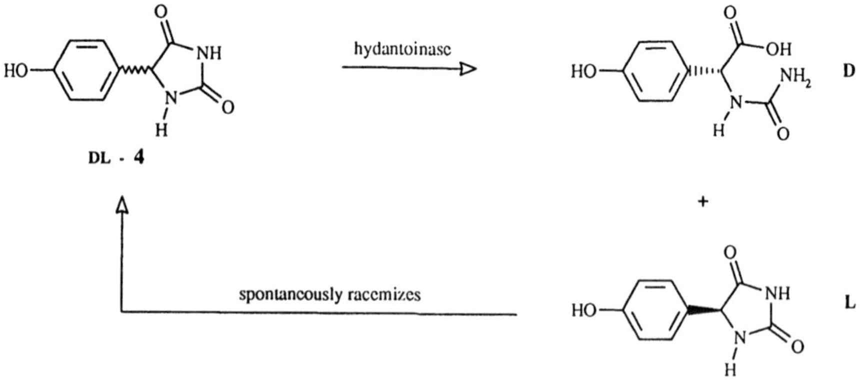

Fig. 4.13. Kinetic resolution with spontaneous racemization.

Similarly, α-amino-ϵ-caprolactam is resolved as a nickel complex by preferential crystallization (Fig. 4.14) (18); the ethanolate of the L-complex precipitates from ethanol, while the D-enantiomer spontaneously racemizes in the basic solution and is recycled.

Amines can be resolved also via crystallization of their camphorsulfonate salts (Fig. 4.15). The insoluble 3(S)-amine camphorsulfonate precipitated, while the undesirable enantiomer was racemized in situ by forming an imine catalytically (19). A 91% yield of the pure enantiomer was thus obtained.

My favorite example is the resolution of synthon **5** (Fig. 4.16) (20), used for making the pyrethroid prallethrin. Racemic acetate **5** was hydrolyzed enantiospecifically by a bacterium to give alcohol **6** that had the wrong stereochemistry. It was, therefore, transformed into a tosylate, which was substituted by water with *inversion* to give the desired compound **7**. The remaining acetate, with the desired stereochemistry, was simply hydrolyzed by water to give the same product **7**. Note the spectacular accomplishment: The synthon was resolved with a 99% yield and 99% enantiomeric purity.

4.2.8. Spontaneous Resolution

There is one method of resolution that does not require optically active agents. This method we would call spontaneous resolution, and it may explain how optical activity first arose on earth (or extraterrestrially). Racemic compounds sometimes take on crystalline forms that have a chiral space group, i.e., they are optically active (21). It is this enantiomorphous crystal that provides the asymmetric environment for kinetic resolution to occur. The reaction must, of course, occur in the solid state, as any solution is racemic. One example is shown in Figure 4.17.

4.2.9. Triage

Another case of what could be termed spontaneous resolution is the example of Pasteur's sodium ammonium tartrate. Some compounds crystallize as conglomerates, which is a mixture of crystals of each pure enantiomer. These can then be

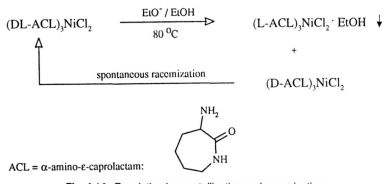

Fig. 4.14. Resolution by crystallization and racemization.

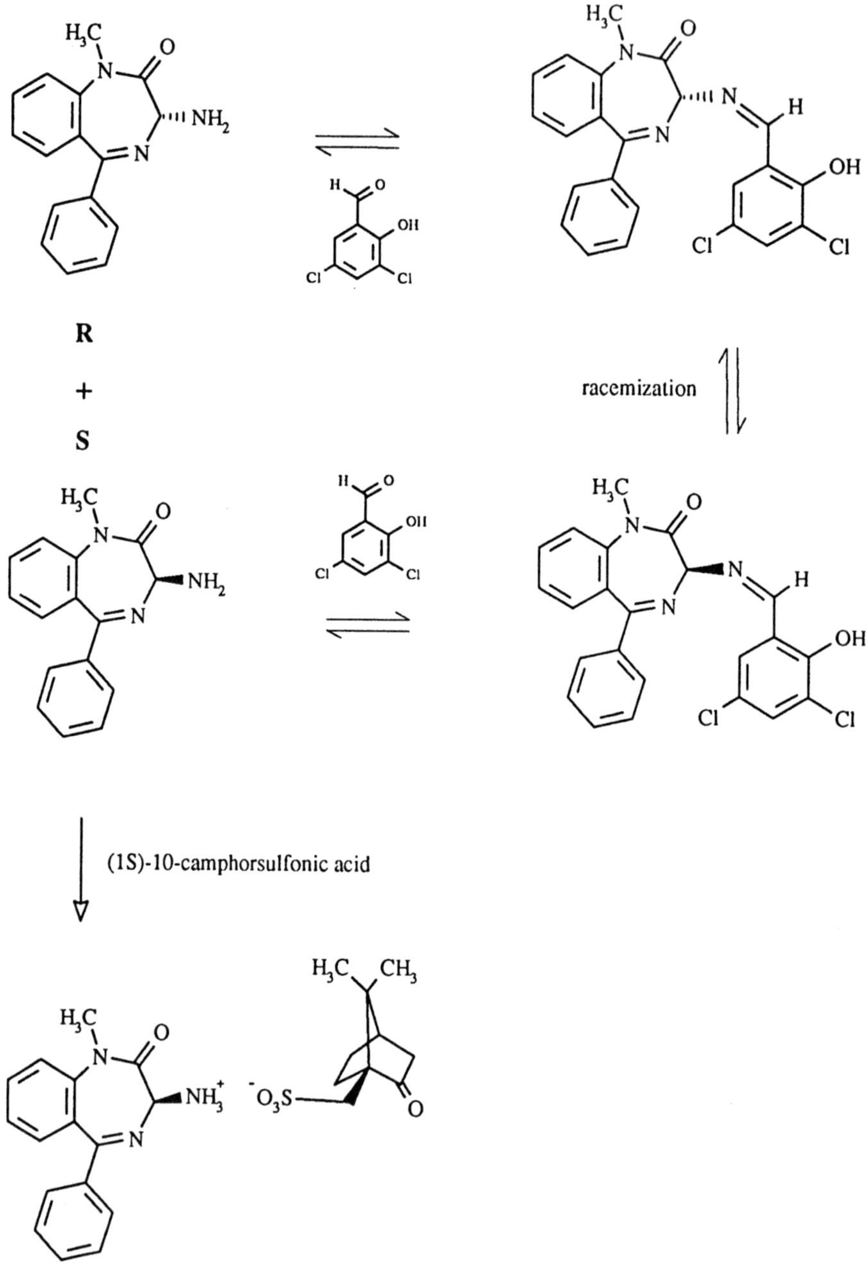

Fig. 4.15. Resolution by crystallization and racemization via imine.

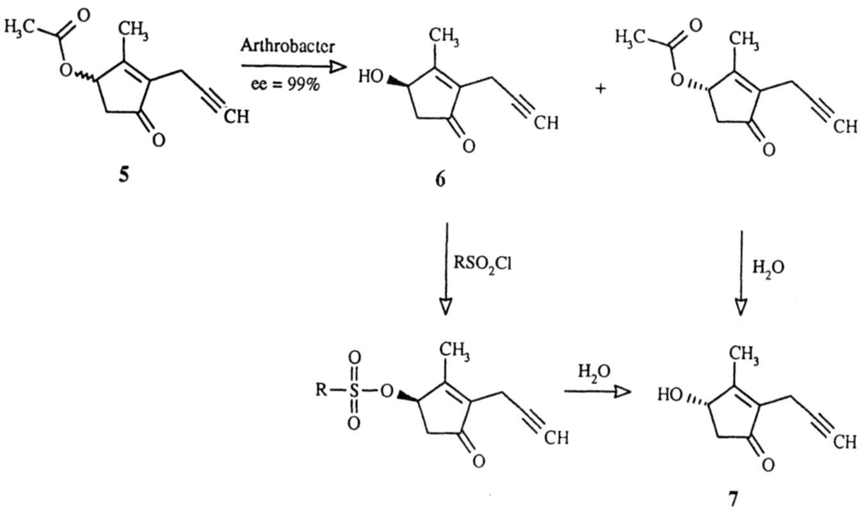

Fig. 4.16. Kinetic resolution with inversion of the distomer.

separated physically, since on careful observation one can differentiate left- and right-handed crystals. However, this procedure is laborious and of little preparative value. Most often this method is used to prepare a few seed crystals to be used in the method of resolution by preferential crystallization, as discussed above.

4.3. ASYMMETRIC SYNTHESIS

Asymmetric synthesis, the second method for obtaining enantioenriched compounds, differs fundamentally from resolution: Asymmetric synthesis starts with achiral compounds (Fig. 4.18) and *creates* chirality (actually, it only transfers it from an enantiopure reagent).

For example, L-phenylalanine, a component of aspartame, is made from achiral cinnamic acid; a lyase adds ammonia to the double bond *enantioselectively* to give L-phenylalanine (Fig. 4.19) (22).

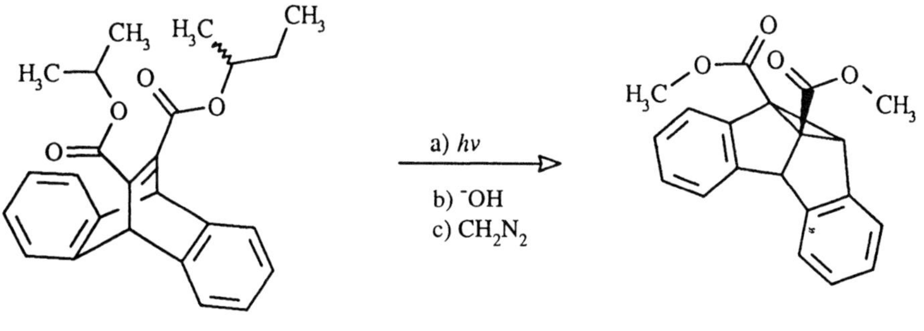

Fig. 4.17. Spontaneous resolution of a dibenzosemibullvalene.

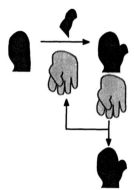

Fig. 4.18. Asymmetric synthesis converts achiral to enantiopure compounds.

Either enantiomer of mevalonolactone (Fig. 4.20) can be synthesized by enzymatic enantioselective hydrolysis of achiral 3-hydroxy-3-methylglutarate (HMG) (23). The resulting acid ester can be treated either with lithium tetrahydridoborate, which reduces the ester to give one enantiomer, or with diborane, which reduces the acid to give the other enantiomer of mevalonolactone. By the way, Lescol, the drug described in Chapter 3, blocks the analogous biochemical reduction by HMG-CoA reductase and thus diminishes the downstream biosynthesis of cholesterol in the body.

An efficient strategy for making enantiopure substances with several stereogenic centers is by desymmetrizing *meso*-compounds (Fig. 4.21). For example, enantiopure lactones can be prepared by microbial oxidation of *meso*-diols (24). Always, the *pro-S* hydroxymethyl group was oxidized preferentially.

The criteria for a good asymmetric synthesis are:

· High optical and chemical yield.
· The enantiopure reagent must be catalytic or else recoverable with high yield and high enantiopurity.
· The product must be separable from the reagent.
· The closer to the reaction site and the more tightly bound the enantiopure reagent, the higher the enantioselectivity will be (this observation explains why asymmetric syntheses with enantiopure solvents do not give high ee values).

More examples of asymmetric syntheses are discussed in Chapters 5 and 6.

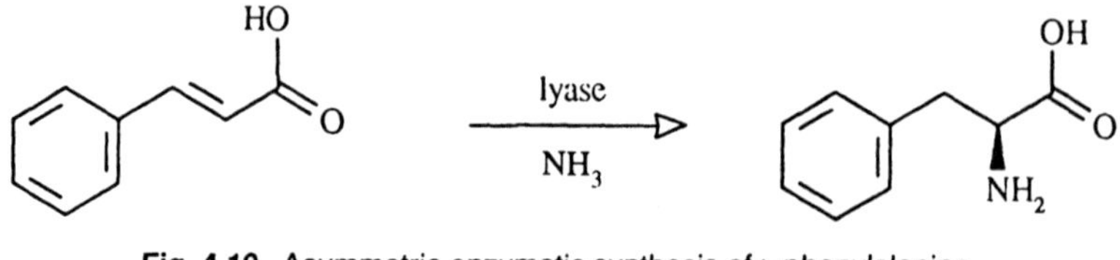

Fig. 4.19. Asymmetric enzymatic synthesis of L-phenylalanine.

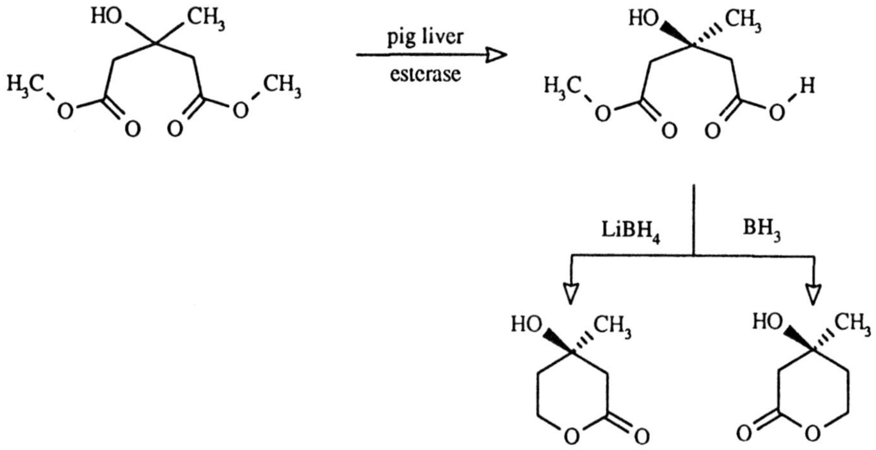

Fig. 4.20. Asymmetric synthesis of both enantiomers of mevalonolactone.

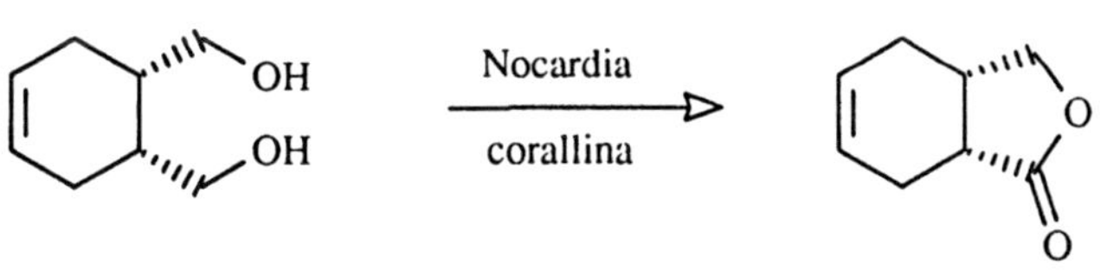

Fig. 4.21. Desymmetrization of a *meso*-compound with a microorganism.

BIBLIOGRAPHY

1. E. L. Eliel and S. H. Wilen, *Stereochemistry of Organic Compounds,* Wiley-Interscience, New York, 1994.

2. C. C. Pfeiffer, "Optical Isomerism and Pharmacological Action, a Generalization," *Science,* **124,** 29–31 (1956).

3. P. A. Lehmann, J. F. Rodrigues de Miranda, and E. J. Ariëns, "Stereoselectivity and Affinity in Molecular Pharmacology," *Prog. Drug Res.,* **20,** 101–142 (1976).

4. H. E. Zaugg, "A Mechanical Resolution of *d,l*-Methadone Base," *J. Am. Chem. Soc.,* **77,** 2910 (1955).

5. T. Watanabe and G. Noyori, "Detailed Operational Conditions of Optical Separation through Fractional Crystallization," *Kogyo Kagaku Zasshi,* **72,** 1083–1086 (1969).

6. N. Sato, T. Uzuki, K. Toi, and T. Akashi, "Direct Resolution of DL-Lysine-3,5-dinitrobenzoate," *Agri. Biol. Chem.,* **33,** 1107–1108 (1969).

7. A. Collet, M. J. Brienne, and J. Jacques, "Optical Resolution by Direct Crystallization of Enantiomer Mixtures," *Chem. Rev.,* **80,** 215–230 (1980).

8. J. Jacques, A. Collet, and S. H. Wilen, *Enantiomers, Racemates and Resolutions,* Wiley-Interscience, New York, 1981.

9. Ger. Pat. DE 1,807,495 (1970), "Separation of the Enantiomers of β-Lactams," H. Jensen (to Farbwerke Hoechst A.-G.).

10. M. B. Groen, H. Schadenberg, and H. Wynberg, "Synthesis and Resolution of Some Heterohelicenes," *J. Org. Chem.*, **36**, 2797–2809 (1971).

11. T. Shibata, I. Okamoto, and K. Ishii, "Chromatographic Optical Resolution on Polysaccharides and Their Derivatives," *J. Liquid Chromatogr.*, **9**, 313–340 (1986).

12. R. A. Sheldon, P. A. Porskamp, and W. ten Hoeve, "Advantages and Limitations of Chemical Optical Resolution," in J. Tramper, H. C. van der Plas, and P. Linko, eds., *Biocatalysts in Organic Syntheses*, Elsevier, Amsterdam, 1985, pp. 59–80.

13. D. Xu, P. G. Mattner, A. Kucerovy, K. Prasad, and O. Repič, "An Efficient Synthesis of Enantiopure SDZ 267-489 via a Resolution/Racemization Method," *Tetrahedron Asymm.*, **7**, 747–754 (1996).

14. W. T. Borden and E. J. Corey, "Synthesis of Optically Active 1,3-di-*t*-Butylallene from Propargylic Derivatives by Hydride Displacement," *Tetrahedron Lett.*, **10**, 313–316 (1969).

15. R. B. Woodward, T. P. Kohman, and G. C. Harris, "A New Optically Active Reagent for Carbonyl Compounds. The Resolution of *d,l*-Camphor," *J. Am. Chem. Soc.*, **63**, 120–124 (1941).

16. U.S. Pat. 3,971,700 (July 27, 1976), "D-Phenylglycinamide and L-Phenylglycine Preparation from DL-Phenylglycinamide by Enzymatic Hydrolysis," W. H. J. Boesten (to Stamicarbon BV).

17. C. J. Sih, and S. H. Wu, "Resolution of Enantiomers via Biocatalysis," *Topics Stereochem.*, **19**, 63–125 (1989).

18. U.S. Pat. 3,988,320 (Oct. 26, 1976), "Resolution—Racemization of α-Aminocaprolactam through Crystallizing Nickel Chloride Complex with Base," S. Sifniades, W. J. Boyle Jr., and F. Van Peppen (to Allied Chemical).

19. P. J. Reider, P. Davis, D. L. Hughes, E. J. J. Grabowski, "Crystallization-Induced Asymmetric Transformation: Stereospecific Synthesis of a Potent Peripheral CCK Antagonist," *J. Org. Chem.*, **52**, 955–957 (1987).

20. H. Hirohara, S. Mitsuda, E. Ando, and R. Komaki, "Enzymatic Preparation of Optically Active Alcohols Related to Synthetic Pyrethroid Insecticides," in Ref. 12, pp. 119–134.

21. M. Garcia-Garibay, J. R. Scheffer, J. Trotter, and F. Wireko, "Generation of Optical Activity Through Solid State Reaction of a Racemic Mixture that Crystallizes in a Chiral Space Group," *Tetrahedron Lett.*, **28**, 4789–4792 (1987).

22. U.S. Pat. 4,598,047 (July 1, 1986), "Phenylalanine Ammonia Lyase-Producing Microbial Cells and Use in Production of Phenylalanine," J. C. McGuire (to Genex Corp.).

23. C. J. Sih and co-workers, "Preparation of (*R*)- and (*S*)-Mevalonic Acids," *J. Am. Chem. Soc.*, **97**, 4144–4145 (1975).

24. H. Luna, K. Prasad, and O. Repič, "Oxidation of Diols to Lactones by *Nocardia corallina* B-276," *Tetrahedron Asym.*, **5**, 303–306 (1994).

FURTHER READING

Y. Izumi and A. Tai, *Stereodifferentiating Reactions,* Academic Press, New York, 1977.

Roger A. Sheldon, *Chirotechnology,* Marcel Dekker, New York, 1993.

5

MIRRORS*

Phospholipids **1** and **2** (Fig. 5.1), analogs of platelet-activating factor (PAF), are related to each other as mirror images; in other words, they are enantiomers. This example is especially interesting, because the enantiomers have not only different activities but also different pharmacological indications (1). The terms *distomer* and *eutomer* are, therefore, confusing here. Drug substance **1** is of interest in the area of oncology (2): It activates macrophages and makes them devour cancer cells. A movie was made that recorded this process, but the biochemical mechanism of the activation is not known. Compound **2** is of interest in the area of multiple sclerosis.

The two differ only in the third dimension: The octadecyl ether in compound **1** is in the front of the plane of the page and the choline phosphate is in the back, whereas the side chains are reversed in **2**. How would one synthesize such compounds? If one formally (retrosynthetically) takes off the side chains, one obtains an achiral tetrahydrofuran-2,2-dimethanol **3**. As we learned in the previous chapter, it is possible to convert achiral compounds to enantiopure products by asymmetric synthesis. The two hydroxyl groups are enantiotopic to each other, and the best reagents for differentiating enantiotopic groups are enzymes.

5.1. FIRST PROCESS

Our first approach to this molecule was, therefore, based on an enzymatic enantiotopos-differentiating reaction (3,4). According to Klibanov (5,6), and our own observations, the best way to accomplish such a differentiation is by hydrolysis of a *bis*-butyrate, using lipases (Fig. 5.2).

*Based on a lecture I gave at the Third French American Chemical Society Meeting, Aussois, France, July 17, 1992.

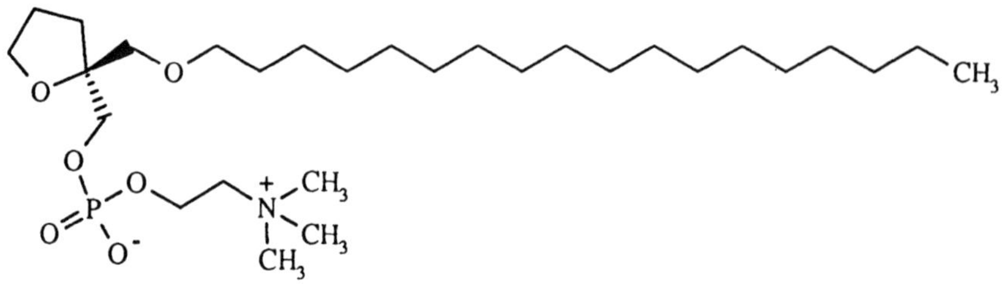

1

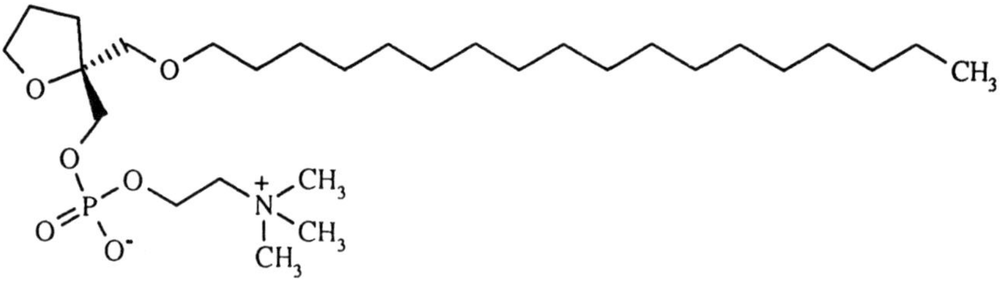

2

Fig. 5.1. Structures of platelet-activating factor analogs.

To this end, we needed to prepare compound **4**, which was accomplished as follows. Commercially available tetrahydrofuran-2-carboxylic acid **5** was esterified with 2,2-dimethoxypropane in acidified methanol to give methyl ester **6**. In a complex step, methyl ester **6** was converted to diol **3**. This conversion involved four chemical reactions: First, diisobutylaluminum hydride (DIBAL) was used to reduce the ester to an aldehyde; second, sodium hydroxide was added to form an enolate; third, the enolate condensed with formaldehyde to form one hydroxymethyl group; and fourth, under these reaction conditions, a Cannizzaro reaction occurred in which the aldehyde was reduced to give the second hydroxymethyl group. The yield of this step was only 60%, mainly because the diol was water soluble and was difficult to isolate. Diol **3** was acylated with 2 moles of butanoic acid chloride to give the required *bis*-butyrate **4**. One enantiotopic butyl ester was differentially hydrolyzed to the *mono*-butyrate **7**. In our hands, the best conditions identified for this conversion were porcine pancreas lipase, pH 7, hexane : water (1 : 1), at room temperature (7). In 45 min, a 89% yield of enantioenriched **7** (ee = 96%) was obtained. The enantiomeric purity of this alcohol, and of the other alcohols later in the synthesis, was determined by ^{31}P NMR using an enantiopure diazaphospholidine derivatization reagent (8). Such good results were obtained only after some optimization, by varying the enzyme and observing the time, yield, and enantiomeric excess of the reaction (Table 5.1).

At first the results were not encouraging; having tried many different lipases (Table 5.1 is only an excerpt), we found the enantiomeric excess of the product was

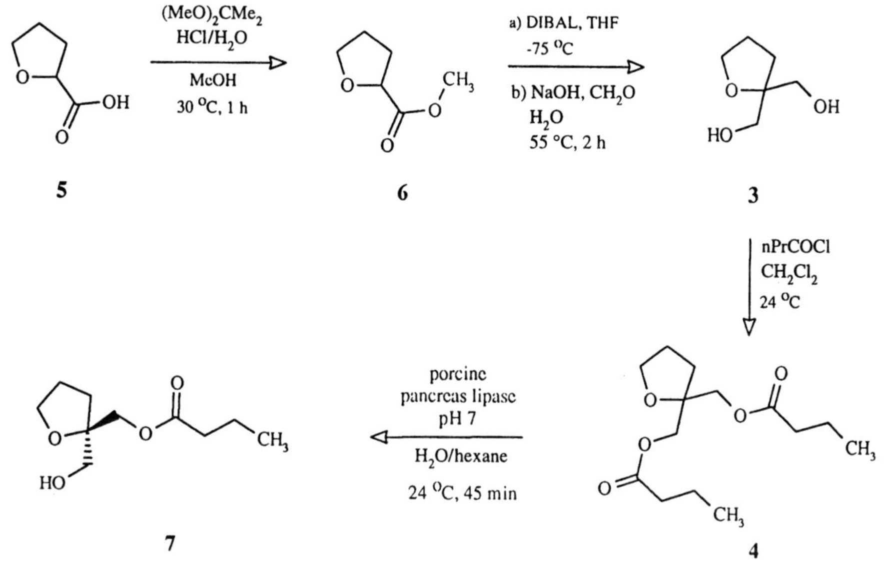

Fig. 5.2. Synthesis of **1** or **2** by enzymatic process (part I).

disappointingly low, perhaps because the long reaction times caused some nonselective hydrolysis or else migration of the butyrate to the other hydroxyl group, a process that causes racemization. To avoid both side reactions, we thought of adding some organic solvent, e.g., 50% of hexane. The intent was to move the product from the aqueous layer into the organic layer and thus protect it from both side reactions (which are faster in water). The result was surprisingly good (99% ee). Even the reaction times were reduced, presumably because an efficient emulsion formed; lipases are known to react at the interface of fat and water. The experiment is conceptually different from Klibanov's conditions in that he uses almost 100% of an organic solvent. When the experiment was upscaled to kilogram scale, the enantiomeric excess went down slightly but was still a practical 94%. The enantiopurity was increased later in the synthesis by crystallization.

Since this was the first time the compound had been made in an enantioenriched form and could not be correlated with any known compounds, we determined its absolute configuration by X-ray crystallography. This technique required appropriate crystals, and several derivatives of **7** were prepared while searching for suitable

Table 5.1. Optimization of Enzymatic Hydrolysis Reaction

Lipase From	Time, h	Yield, %	ee, %	Comment
Candida cylindracea	2	54	48	
Porcine liver	1.5	35	36	
Mucor javanicus	5.5	64	73	
Porcine pancreas	0.75	89	99	+50% hexane
Porcine pancreas	0.5	90	94	Scaled up

crystals (Fig. 5.3). *Mono*-butyrate **7** was allowed to react with camphorsulfonic acid chloride, then hydrolyzed with methanolic potassium carbonate, but the best crystals were obtained as the double derivative of camphorsulfonate and *p*-nitrobenzoate **8**. The X-ray analysis indicated that the compound had the *S* configuration, so all figures show the correct absolute stereochemistry (7).

The synthesis continued (Fig. 5.4) by protecting the rear hydroxyl group in **7** with the trityl-protecting group and by removing the butyrate in **9** with methanolic potassium carbonate. The front hydroxyl group in **10** was then allowed to react with octadecyl bromide, catalyzed by tetra-*n*-butylammonium iodide, to give the octadecyl ether **11**.

The trityl protecting group was removed (Fig. 5.5) with trifluoroacetic acid to give **12**. Compound **12** was crystallized from heptane, by which the enantiomeric excess was increased to 99%. Choline phosphate was then attached to the exposed rear hydroxyl group. This transformation was performed in a one-pot, three-step sequence: alcohol **12** was allowed to react with phosphorus trichloride for 2 h in the presence of triethylamine, the second chloride was allowed to react with choline tosylate in the presence of pyridine at 50°C for 5 h, and the third chloride was hydrolyzed by water at 55°C for 5 h. Finally, the crude product was passed through a mixed-bed ion exchange column (Amberlite MB-1) that removed all acids and bases in the mixture and allowed only the neutral, zwitterionic product **1** to pass

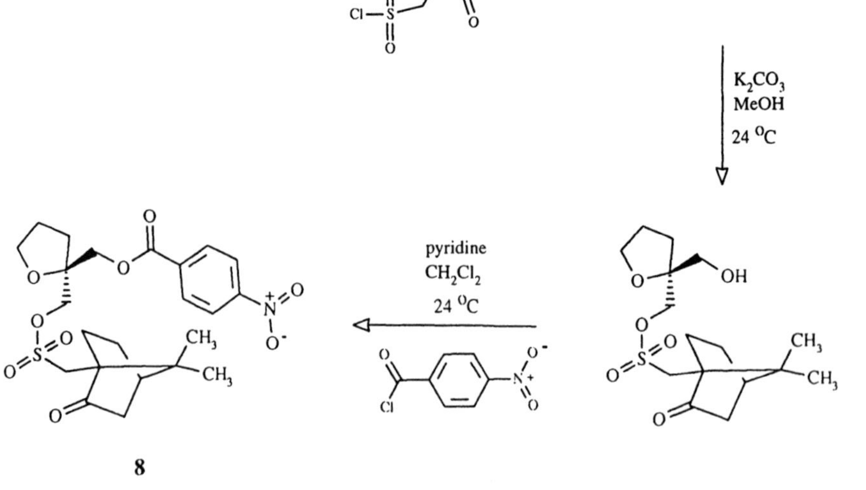

Fig. 5.3. Derivatives for X-ray structure determination.

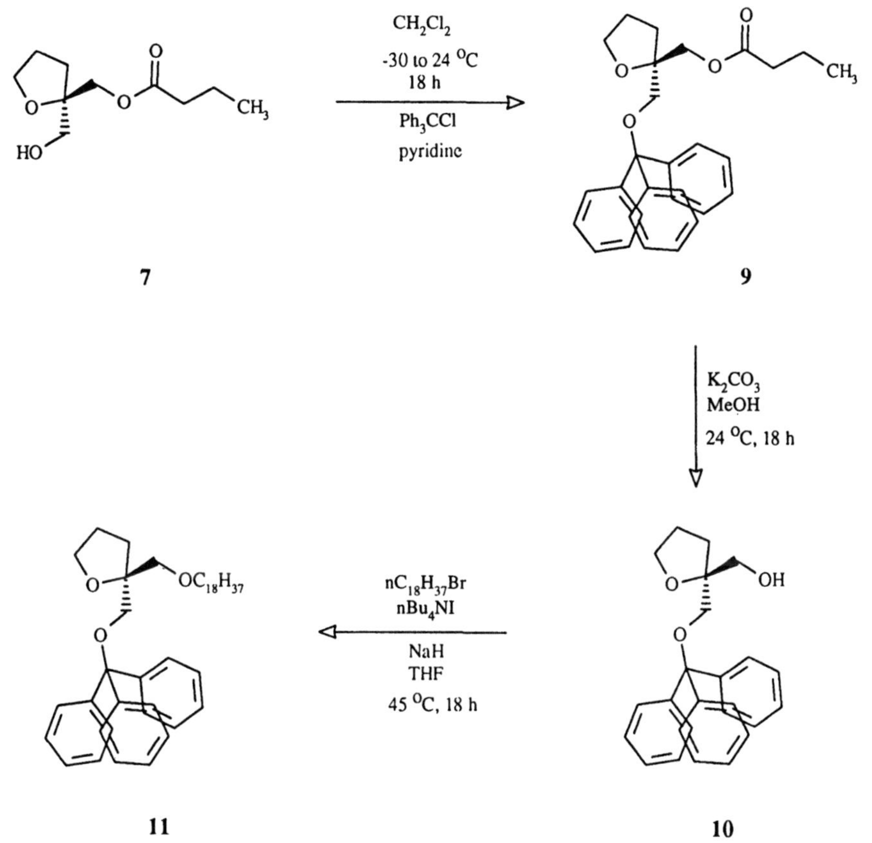

Fig. 5.4. Synthesis of **1** by enzymatic process (part II).

through. The complex sequence of reactions was sufficiently optimized so that the ion exchange treatment was the only purification needed, and the drug substance was simply isolated by lyophilization. Enantiomer **1** had a specific rotation of $[\alpha]_D - 1.9$ ($c = 1.0$, ethanol). To prepare a more crystalline and less hygroscopic drug substance, several crystallization methods were tried (isopropanol–acetone, ethanol–acetone, tetrahydrofuran–acetone, tetrahydrofuran–ethanol). None had superior properties to the freeze-dried material.

Enzymes have one disadvantage: There is only one form of the enzyme. As here, if both enantiomers of the drug substance are needed, how does one make the other enantiomer? Leave it to the ingenuity of organic chemists. The other enantiomer **2** was prepared *from the same intermediate* **10** (Fig. 5.6): the front hydroxyl group was protected by a second protecting group, e.g., a benzyl ether (**10 → 13**); the rear protecting group was removed by trifluoroacetic acid to give **14**, allowing the octadecyl side chain to be placed on the rear hydroxyl group (**14 → 15**).

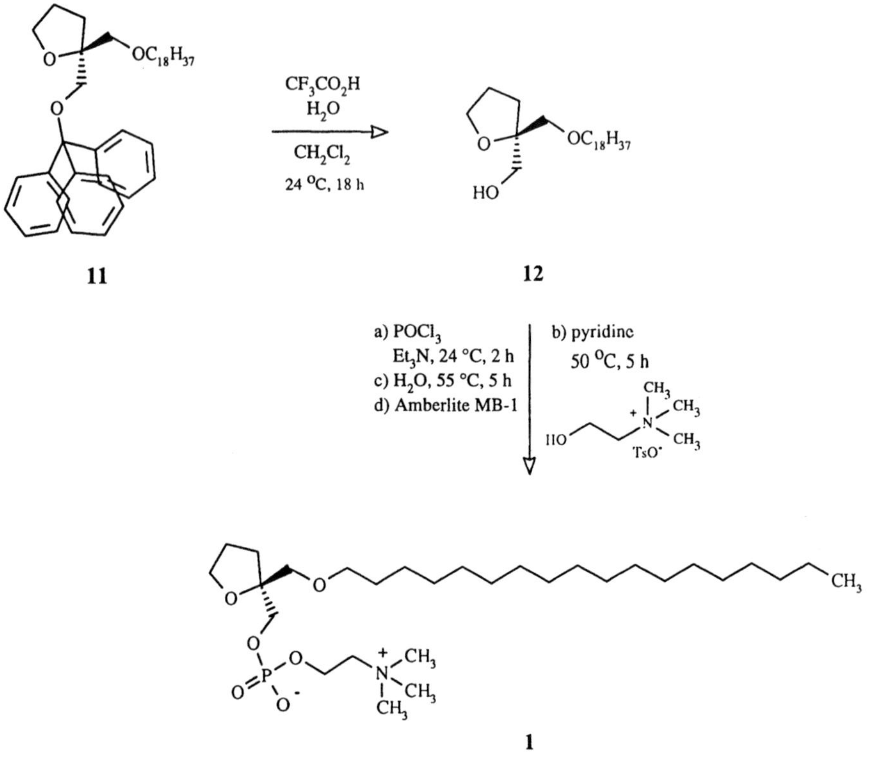

Fig. 5.5. Synthesis of **1** by enzymatic process (end).

The benzyl-protecting group was removed by hydrogenation (Pd/C at 48 psi of hydrogen gas) to give the key precursor **16** (Fig. 5.7), which was recrystallized from heptane to increase the enantiomeric excess to 99%. Choline phosphate was attached to the exposed front hydroxyl group as described before for the other enantiomer in Fig. 5.5. Enantiomer **2** had a specific rotation of $[\alpha]_D + 1.9$ ($c = 1.0$, ethanol).

The goal of this research was thus achieved: Using an original synthetic design, process development laboratories have, for the first time, prepared the two enantiomers of this compound, each a drug candidate. However, this synthesis was not an ideal process mainly because it was too long: 9 steps to the (−)-enantiomer and 11 steps to the (+)-enantiomer, of which 6 steps were common to both syntheses. At least 4 of the 11 steps were protection or deprotection steps, which is not a sign of an efficient synthesis (9).

5.2. SECOND PROCESS

Consequently, we looked for other, more efficient pathways to compounds **1** and **2**. A publication by Harada and co-workers (10) gave us the idea for this approach.

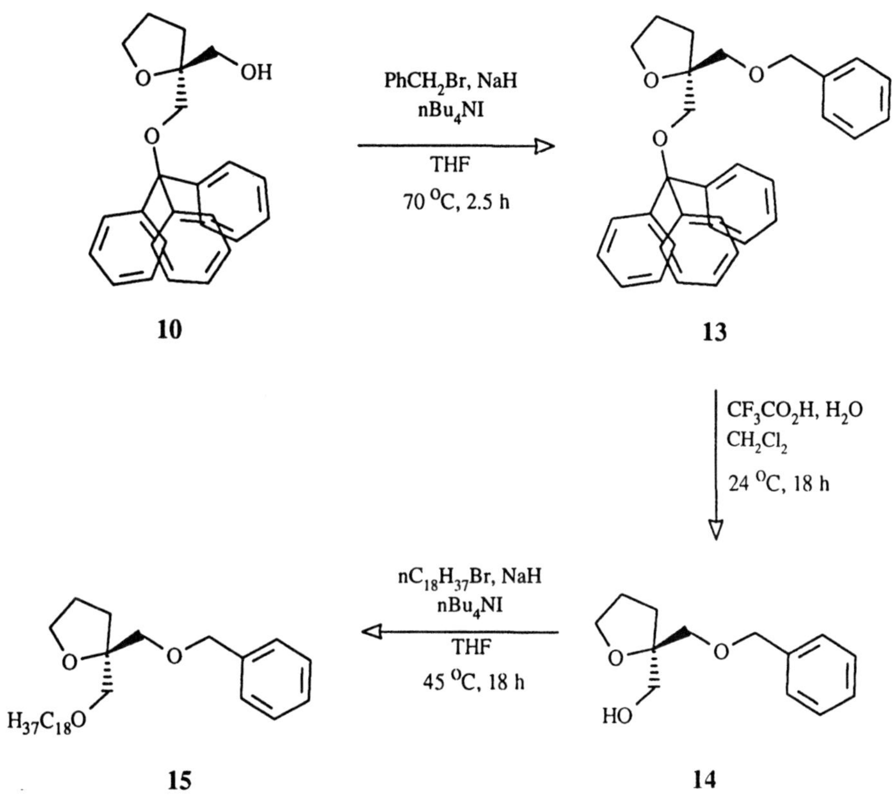

Fig. 5.6. Synthesis of **2** by enzymatic process (part II).

They used *l*-menthone as the asymmetric environment to differentiate two enantiotopic hydroxyl groups.

Similarly, we silylated tetrahydrofuran-2,2-dimethanol **3** with trimethylsilyl trifluoromethanesulfonate to give **18** (Fig. 5.8), and, in the same pot at −40°C, combined it with *l*-menthone (**17**) to give a 1:3 mixture of diastereomeric ketals **19** and **20** (11). This ratio could not be improved since the reaction at lower temperatures was much slower and gave lower yields. The two diastereomers were separated by silica gel chromatography. Cyclic ketal **20** was then opened (Fig. 5.9) with titanium chloride at −90°C, presumably forming an oxonium species to which allyltrimethylsilane was added. Both reactions occurred exclusively at the *equatorial* position, forming only isomer **21**. This enabled the differentiation of the enantiotopic hydroxyl groups. The octadecyl ether could thus be attached to the front hydroxyl (**21 → 22**), and the transient synthon was removed with trifluoroacetic acid to give **12**, which showed an enantiomeric excess of 99% even before crystallization. Compound **12** was converted to drug substance **1** in one pot, as described above and in Fig. 5.5.

The other enantiomer can be made in the same way from the other diastereomeric ketal **19**. If the amount of **19** is insufficient (since it is isolated with a yield

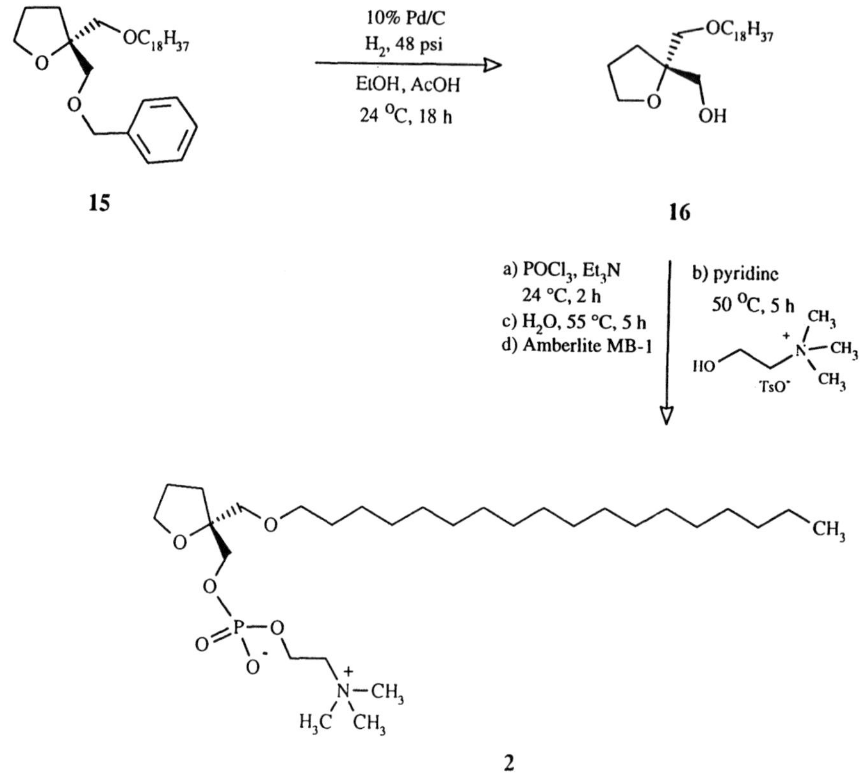

15 → **16**

reagents over arrow: 10% Pd/C, H₂, 48 psi, EtOH, AcOH, 24 °C, 18 h

a) POCl₃, Et₃N 24 °C, 2 h
c) H₂O, 55 °C, 5 h
d) Amberlite MB-1

b) pyridine 50 °C, 5 h

2

Fig. 5.7. Synthesis of **2** by enzymatic process (end).

of only 20%), enantiomer **2** can be made also from the common intermediate **21** (Fig. 5.10). As was the strategy in the enzymatic process, the front hydroxyl was protected as a benzyl ether **23**, the chiral auxiliary was removed with trifluoroacetic acid to give **14**, and the rear hydroxyl was alkylated with octadecyl bromide leading to **15**. Intermediate **15** was converted in two steps into the (+)-enantiomer of the drug substance **2**, as already described above and in Fig. 5.7.

This second enantioselective synthesis produced either enantiomer **1** or **2** in excellent purity (ee > 99%) and in only five and seven steps, respectively. Although this process was the shortest and purest synthesis so far, it had two disadvantages: Chromatography was needed to separate the diastereomeric menthone ketals **19** and **20**, and the chiral auxiliary (menthone) could not be recycled because it was reduced in the process.

5.3. THIRD PROCESS

For these reasons, we designed yet a third enantioselective synthesis, which used the classical Sharpless asymmetric epoxidation as the enantioselective step (12,13).

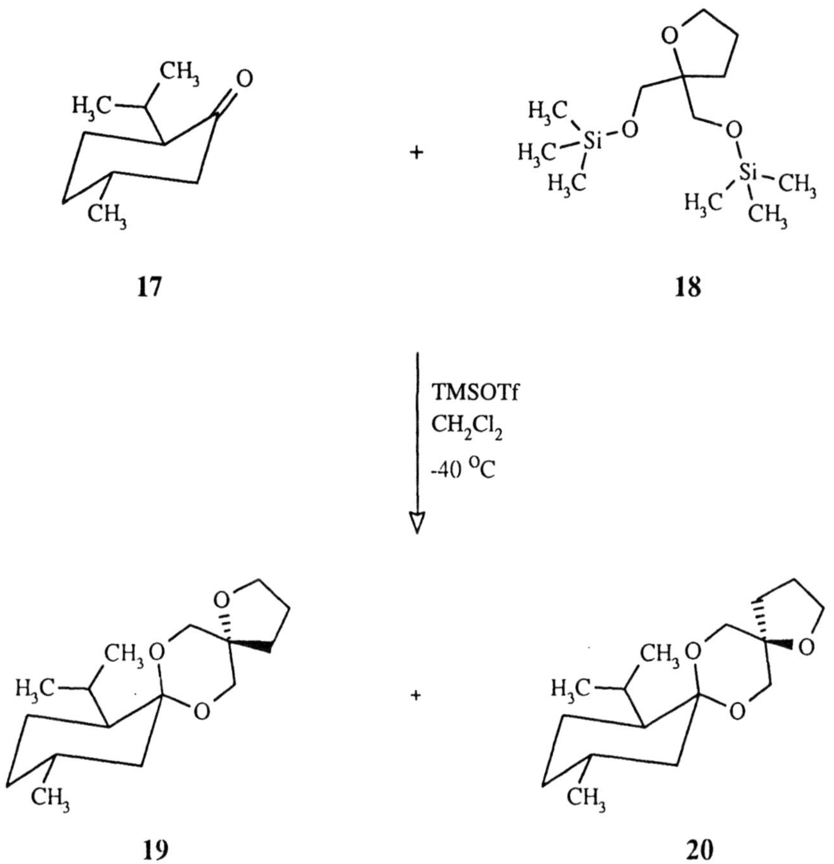

17 + **18**

TMSOTf
CH₂Cl₂
-40 °C

19 + **20**

Fig. 5.8. Synthesis of **1** or **2** by *l*-menthone route (part I).

It is not immediately obvious how an epoxide could serve as a synthon in a synthesis of a tetrahydrofuran, but in Fig. 5.11 we see that a propanol tail can open the enantiopure epoxide to give the key precursor **16** directly. To prepare for this dramatic reaction, we first needed the appropriate allyl alcohol, the necessary substrate for a Sharpless epoxidation (14). 3-Bromopropanol was protected as a ketal with 2-methoxypropene in the presence of pyridinium *p*-toluenesulfonate (Fig. 5.12). The bromine in **24** was substituted with malonate to give **25**. The diester **25** was reduced to the allyl alcohol **26** in a one-pot series of reactions: Sodium hydride formed an enolate of the ester, which was reduced in a conjugate manner with diisobutylaluminum hydride, stopping at the allyl alcohol stage **26**. There was little of the saturated alcohol produced.

The enantioselective Sharpless epoxidation (*t*-butyl hydrogen peroxide, (+)-diethyl tartrate, titanium tetraisopropoxide, molecular sieves, in dichloromethane at −20°C for 3 h) produced the desired epoxide **27** with an ee of about 92% (Fig. 5.13). At first we feared that compound **27** may racemize in the next reaction via a Payne rearrange-

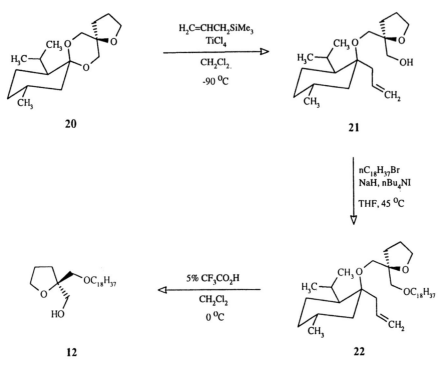

Fig. 5.9. Synthesis of **1** by *l*-menthone route (part II).

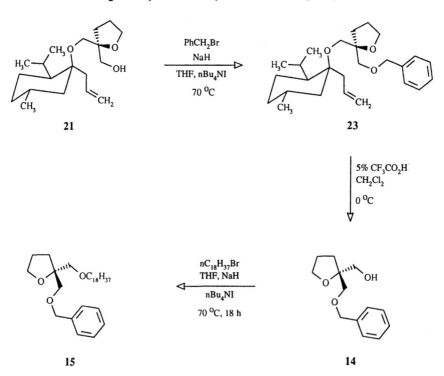

Fig. 5.10. Synthesis of **2** by *l*-menthone route (part II).

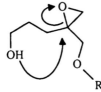

Fig. 5.11. Epoxide as synthon for enantiopure tetrahydrofurans.

ment. Professor Sharpless, however, pointed out to us that a Payne rearrangement of 2-substituted glycidols like **27** is degenerate, i.e., the rearrangement produces the same molecule. On second thought, this result is clear: Although the Payne rearrangement inverts the stereochemistry at the stereogenic carbon, it also interchanges the functional groups, resulting in no total change (15). Anyway, isotope labeling has shown that no Payne rearrangement occurs in this nonprotic reaction (16).

Alkylation of the free hydroxyl group thus gave the octadecyl ether **28**. In the key step, the ketal protecting group was removed with pyridinium *p*-toluenesulfonate in acetic acid; simultaneously, the resulting hydroxyl group cyclized, opening the epoxide and forming the tetrahydrofuran ring. Indeed, this molecule is **16**, the precursor to the drug substance; **16** was crystallized to an enantiomeric excess of 99% and converted to the drug substance **2** in one step, as in previous syntheses (Fig. 5.7). Enantiomer **1** was prepared *by the same process* just by replacing (+)-diethyl tartrate with (−)-diethyl tartrate.

Though this third route to **1** and **2** was innovative and efficient (either enantiomer can be prepared by the same 7-step process simply by choosing the appropriate

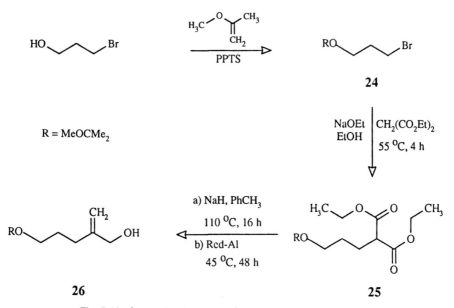

Fig. 5.12. Synthesis of **1** or **2** via Sharpless epoxidation (part I).

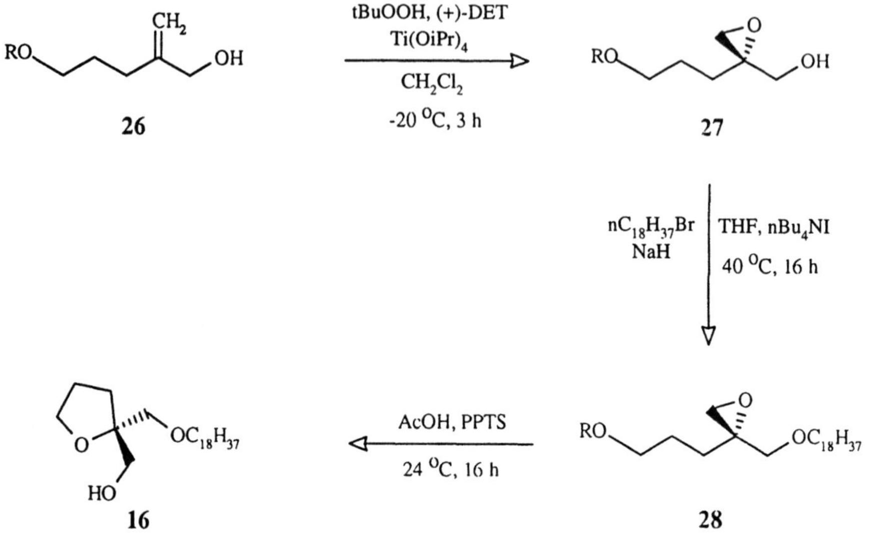

Fig. 5.13. Synthesis of **2** via Sharpless epoxidation (part II).

enantiomer of diethyl tartrate), it had four disadvantages that made it a nonideal process. (*1*) The Sharpless epoxidation required the use of *t*-butyl hydrogen peroxide, which our production department considered too hazardous to store in large quantities. (*2*) The reaction also required the use of dichloromethane as a solvent and was, therefore, an ecological problem. (*3*) The reaction used large quantities of molecular sieves, which could not be recycled (due to the presence of peroxide) and were a disposal problem. (*4*) Finally, and most important, the yield of the synthesis was low and variable, perhaps due to the labile ketal protecting group. For these reasons, this process was abandoned, and the original enzymatic route revisited.

5.4. FOURTH PROCESS

Above we noted that the only problem with the first process was that it had too many steps, mostly too many protection–deprotection steps. This objection was eliminated by shortening the synthesis. The first half of the synthesis was shortened by one step by avoiding the isolation of the diol **3** (Fig. 5.14).

The added benefit was an increased yield; the diol was water soluble and previously difficult to isolate. The new process effected the conversion of tetrahydrofuroate **6** to the *bis*-butyrate **4** in one pot, this time using butanoic anhydride as the acylating agent. Compound **4** was obtained with a 70% yield and 98% purity. The synthesis was shortened by another step by replacing the trityl protecting group, which had been difficult to remove, with a ketal (Fig. 5.15). This ketal protecting group is similar to the third process above, but now it is needed only in one step, so its labile nature is an advantage: It is removed in acidic workup, not in a separate

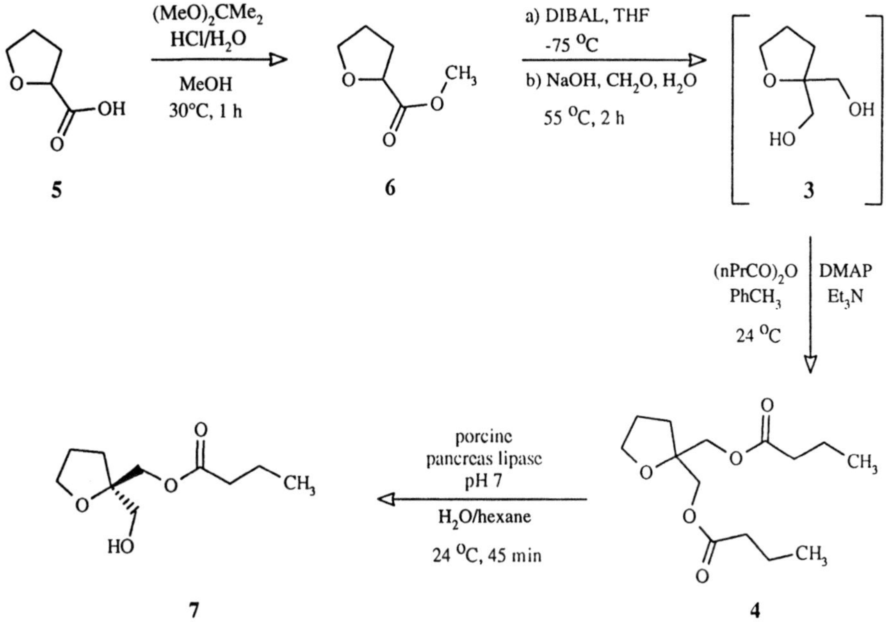

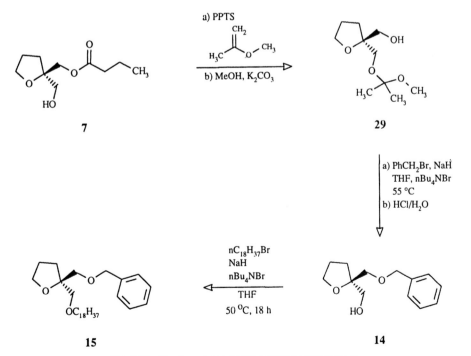

Fig. 5.14. Shortened enzymatic synthesis of **1** or **2** (part I).

Fig. 5.15. Shortened enzymatic synthesis of **2** (part II).

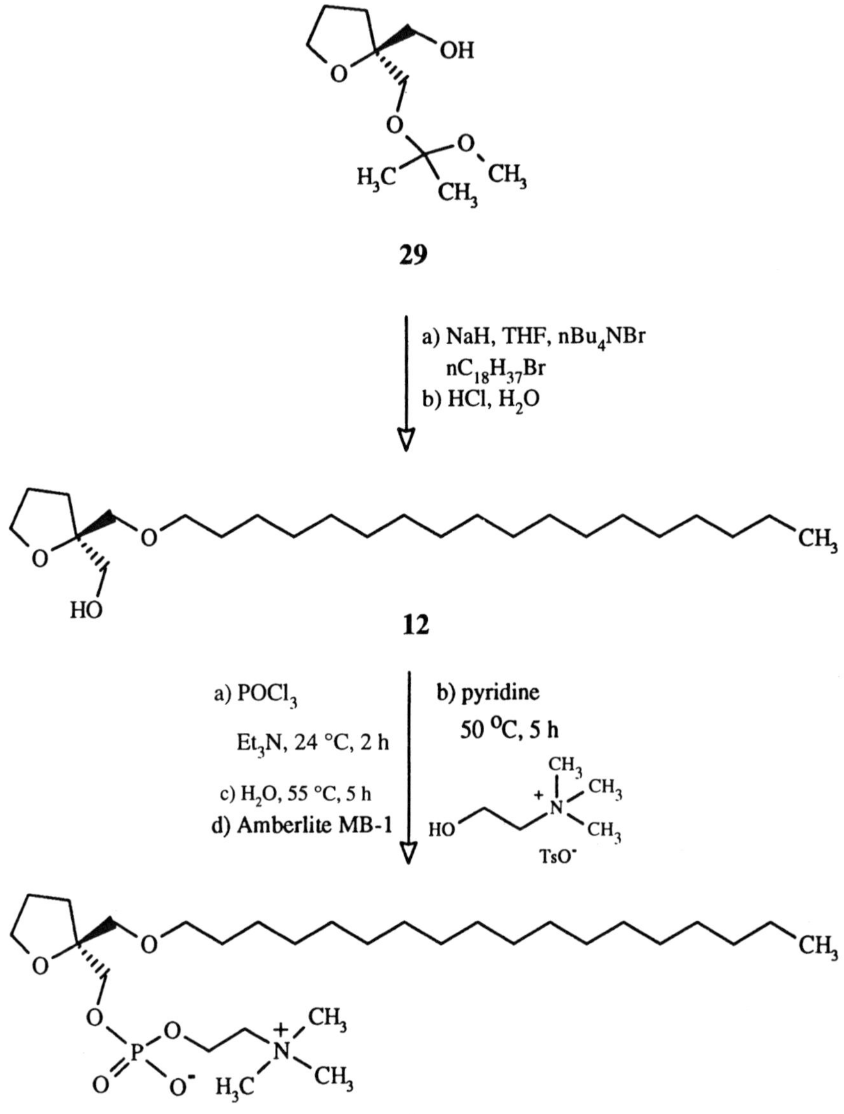

29

a) NaH, THF, nBu₄NBr
   nC₁₈H₃₇Br
b) HCl, H₂O

12

a) POCl₃

   Et₃N, 24 °C, 2 h

c) H₂O, 55 °C, 5 h
d) Amberlite MB-1

b) pyridine
   50 °C, 5 h

1

Fig. 5.16. Shortened enzymatic synthesis of **1** (end).

step. The difference is not only semantic as this protecting group really reduces the number of plant operations. As we have seen before (Fig. 5.7), intermediate **15** was converted into **2** in two steps, so the synthesis has a total of eight steps. On the other hand, intermediate **29** was converted (Fig. 5.16) into drug substance **1** in only two steps, for a total of six steps. This synthesis was successfully upscaled in our pilot plant, and several kilograms of these drug substances were prepared.

This example illustrates the great power and efficiency of *asymmetric synthesis*: The enantioselective enzymatic hydrolysis of an achiral intermediate, with a theoretical yield of 100%, produced the enantioenriched product in a yield of 90%. This approach was more efficient and conceptually quite different from an enzymatic kinetic *resolution* (same reaction conditions are used) of a racemic intermediate, with a theoretical yield of 50% and an actual yield of 32% (see Fig. 4.10).

Mirrors are magic after all.

BIBLIOGRAPHY

1. Eur. Pat. 462 935 (1992), "New *R*-Enantiomer of 2-Tetrahydrofuran-2-yl-phosphinyloxy Derivatives, for Treatment of Tumors and Multiple Sclerosis," H. Estermann, W. Houlihan, K. Prasad, and R. Underwood (to Sandoz).

2. W. J. Houlihan, K. Prasad, R. Underwood, O. Repič, and P. G. Munder, "Antitumor Activity of the *R*- and *S*-Enantiomers of 2-[[Hydroxy[[2-[(octadecyloxy)methyl]tetrahydrofuran-2-yl]methoxy]-phosphinyl]oxy]-*N,N,N*-trimethylethylaminium Hydroxide Inner Salt," *J. Med. Chem.,* **39,** 605–608 (1996).

3. U.S. Pat. 5,229,377 (July 20, 1993), "Process for the Preparation of the *R*-Stereoisomer of the Monobutyrate Ester of 2,2-*bis*(Hydroxymethyl)tetrahydrofuran, Its Use in Preparing Stereoisomers of Pharmacologically Active Compounds, and Certain Specific Stereoisomers Produced Thereby," H. Estermann, K. Prasad, R. Underwood, and W. Houlihan (to Sandoz).

4. K. Prasad, H. Estermann, R. L. Underwood, C.-P. Chen, A. Kucerovy, and O. Repič, "Asymmetrization of Tetrahydrofuran-2,2-dimethanol: Synthesis of the Enantiomers of SRI 62-834," *J. Org. Chem.,* **60,** 7693–7696 (1995).

5. G. Kirchner, M. P. Scollar, and A. M. Klibanov, "Resolution of Racemic Mixtures via Lipase Catalysis in Organic Solvents," *J. Am. Chem. Soc.,* **107,** 7072–7076 (1985).

6. A. M. Klibanov, "Enzymes That Work in Organic Solvents," *ChemTech,* **16,** 354–359 (1986).

7. H. Estermann, K. Prasad, M. J. Shapiro, O. Repič, G. E. Hardtmann, J. J. Bolsterli, and M. D. Walkinshaw, "Enzyme-catalyzed Asymmetrization of *bis*(Hydroxymethyl)-Tetrahydrofurans," *Tetrahedron Lett.,* **31,** 445–448 (1990).

8. A. Alexakis, S. Mutti, and P. Mangeney, "A New Reagent for the Determination of the Optical Purity of Primary, Secondary, and Tertiary Chiral Alcohols and of Thiols," *J. Org. Chem.,* **57,** 1224–1237 (1992).

9. B. M. Trost, "The Atom Economy—A Search for Synthetic Efficiency," *Science,* **254,** 1471–1477 (1991).

10. T. Harada, Y. Ikemura, H. Nakajima, T. Ohnishi, and A. Oku, "Enantiodifferentiating Functionalization of Prochiral Diols by Highly Stereoselective Ring-Cleavage Reaction of Spiroacetals Derived from *l*-Menthone with Allyltrimethylsilane—Titanium Tetrachloride," *Chem. Lett.,* **1990,** 1441–1444.

11. K. Prasad, R. L. Underwood, and O. Repič, "Asymmetrization of Tetrahydrofuran-2,2-dimethanol Using *l*-Menthone as a Chiral Template," *J. Org. Chem.,* **61,** 384–385 (1996).

12. U.S. Pat. 5,208,352 (May 4, 1993), "Process for Preparing the *R*- and *S*-Isomers of 2-Hydroxymethyl-2-octadecyloxymethyltetrahydrofuran and Their Use in Preparing

Stereoisomers of Pharmacologically Active Compounds," C.-P. Chen, K. Prasad, and W. J. Houlihan (to Sandoz).

13. R. A. Johnson and K. B. Sharpless, "Addition Reactions with Formation of Carbon-Oxygen Bonds: Asymmetric Methods of Epoxidation," in B. M. Trost, I. Fleming, and S. V. Ley, eds., *Comprehensive Organic Synthesis,* Pergamon Press, New York, 1991, pp. 389–436.

14. K. Prasad, W. J. Houlihan, C.-P. Chen, R. L. Underwood, and O. Repič, "An Enantioselective Construction of a Tetrahydrofuran Synthon via an Epoxy Alcohol: Synthesis of *S*-Enantiomer of SRI 62-834," *Tetrahedron Asym.,* **7**, 837–842 (1996).

15. G. B. Payne, "Epoxide Migrations with α,β-Epoxy Alcohols," *J. Org. Chem.,* **27**, 3819–3822 (1962).

16. J. Bajwa and U. B. Sunay, "Isotopic Labelling as a Probe in the Study of a Degenerate Payne Rearrangement in a Tertiary Epoxide System," *J. Label. Compounds Radiopharm.,* in press.

6

AU NATUREL*

As the title implies, the chiral pool method will be demonstrated in this chapter (1). This strategy looks to nature (the pool of optically active molecules) to provide enantiomerically pure building blocks from which optically active drugs can be constructed. Hanessian termed these chiral synthons chirons (2); he has even composed a computer program that recognizes features (stereochemistry and functional groups) in a natural product that could become the required parts of the synthetic target.

This chapter is the climax to which the preceding discussions were leading: How to synthesize the cholesterol-lowering agent that was introduced in Chapter 3, in an enantiopure form? It turned out that the (R,S)-enantiomer (the eutomer) was the active compound and that the (S,R)-enantiomer (the distomer) was only weakly active in rats. Three syntheses will be described: The first one starting from D-glucose, the second from L-malic acid, and the third one from L-mandelic acid. The first two are chiral pool strategies, whereas the third one is, strictly speaking, an asymmetric synthesis; the difference is whether the synthon ends as part of the product or whether it is strictly a reagent (and, it is hoped, recyclable).

In Chapter 3 (Fig. 3.7), the racemic synthesis of the six-carbon synthon for the side chain of the cholesterol-lowering agents was described. This side chain contains all the stereogenic atoms in the molecule and is also primarily responsible for the activity of the molecule; the structure of the heterocyclic part of the molecule was then optimized by medicinal chemists to maximize the activity. To make the story more varied, we will attach the side chain to another heterocycle, an imidazole instead of an indole (3).

---

*Based on a lecture I gave at the Sixth European Symposium on Organic Chemistry, Belgrade, Yugoslavia, September 14, 1989.

The synthesis (Fig. 6.1) can be divided into four problems:

· How to synthesize the imidazole portion?
· How to obtain exclusively the *E*-olefin?
· How to obtain stereoselectively the *syn*-1,3-diol?
· How to obtain enantiopure (*R,S*)-1,3-diol?

Since we already discussed the solutions to Questions 2 and 3 for the racemic synthesis in Chapter 3, we will deal here with only problems 1 and 4.

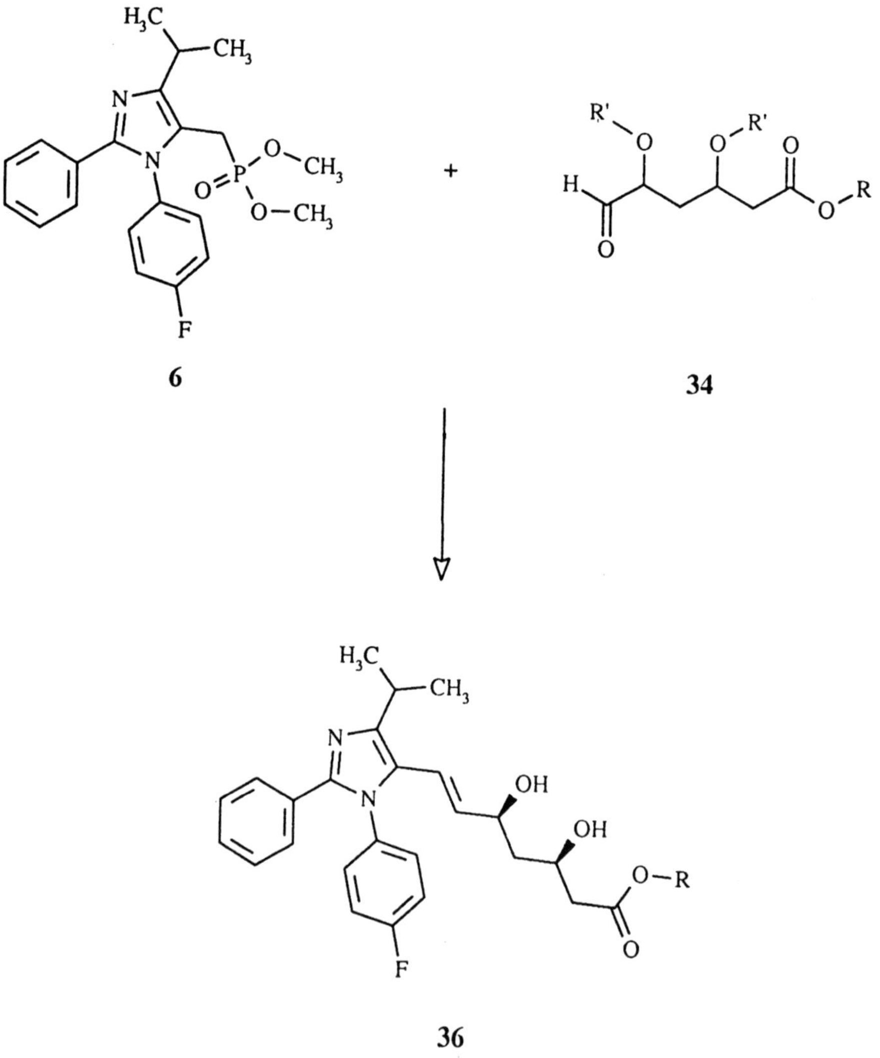

Fig. 6.1. Synthetic strategy for HMG-CoA-reductase inhibitors.

6.1. SYNTHESIS OF THE IMIDAZOLE PORTION

The imidazole in the target molecule **36** was also prepared from a natural product, valine. However, as we will see in a moment, this raw material does not have to be enantiopure, as it will lose its stereochemistry in the third step of the synthesis (Fig. 6.2). Thus DL-valine was benzoylated to introduce the phenyl group **1**. Next, the carboxylic acid group had to be replaced by an acyl group (**1 → 2**). This transformation was accomplished by acylating the amino acid in the α-position using triethylamine–dimethylaminopyridine as the base and acetic anhydride as the acylating agent; the resulting β-keto-acid decarboxylated under the reaction conditions and gave the dicarbonyl compound **2**. The only atoms missing at this point were those of p-fluoroaniline, which was next condensed with the dicarbonyl compound **2**, with the assistance of p-toluenesulfonic acid as the acid catalyst, to give imidazole **3**.

To prepare the heterocycle for the Horner–Emmons reaction (Fig. 6.3), the methyl group was oxidized with potassium persulfate and copper(II) sulfate to aldehyde **4**. The aldehyde was reduced with sodium borohydride to give alcohol **5**. To avoid the reduction step entirely, other methods of oxidation of the benzylic methyl group directly to the alcohol were tried. The most promising reagent was $Mn(OAc)_3$.

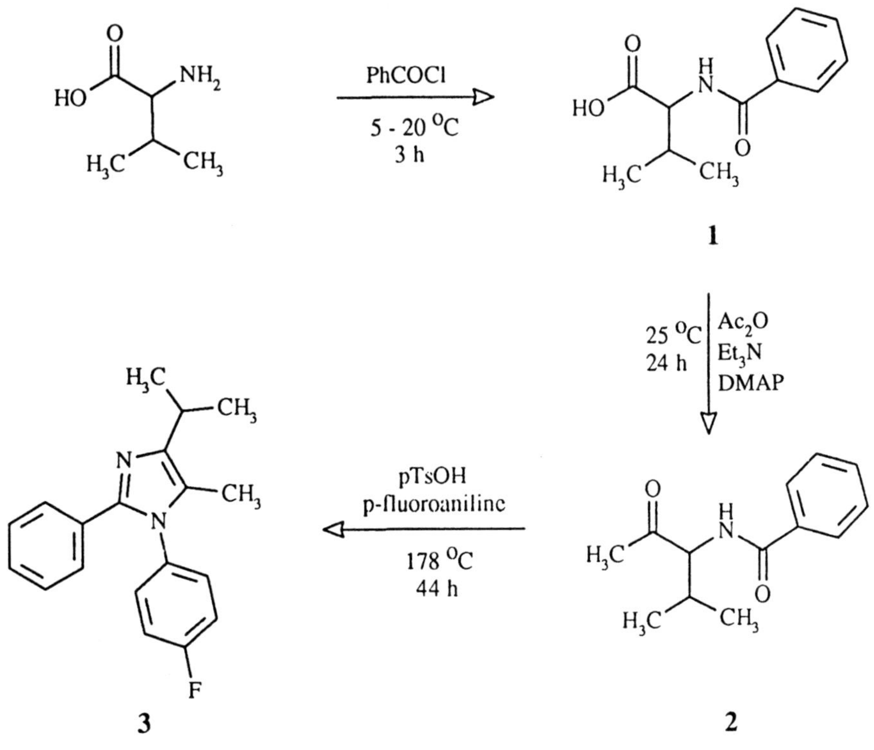

Fig. 6.2. Synthesis of the imidazole portion.

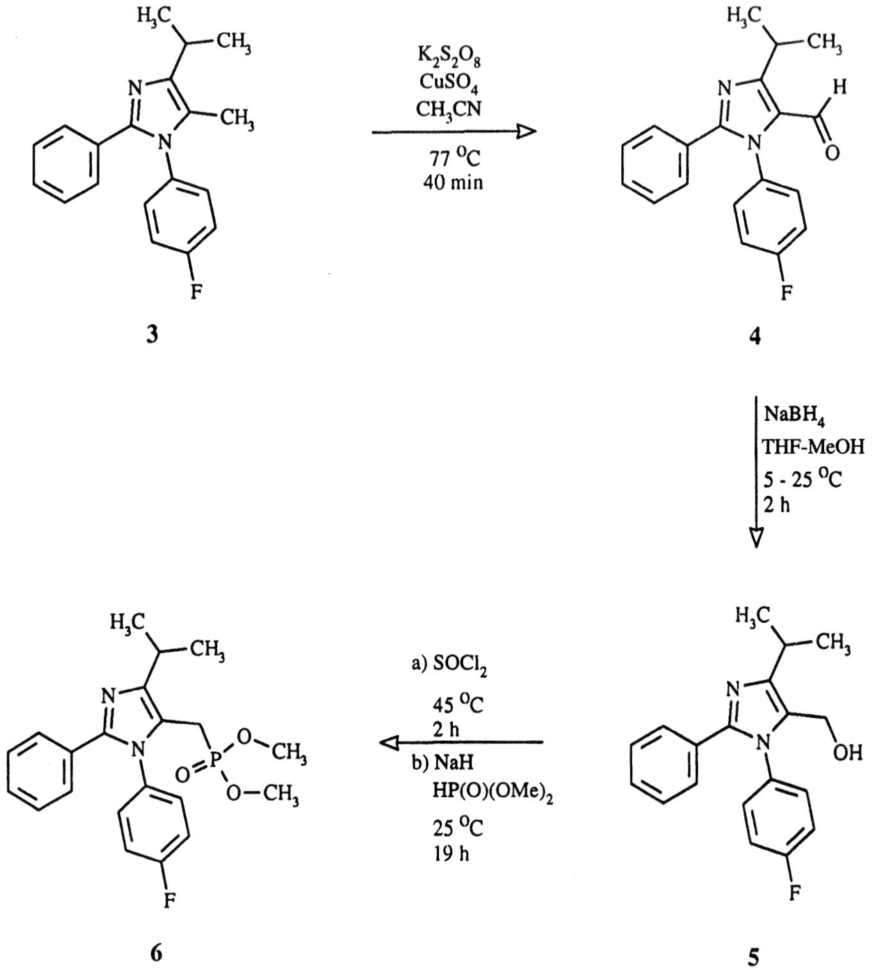

Fig. 6.3. Synthesis of phosphonate intermediate.

Its disadvantage is that the reagent must be prepared from $Mn(OAc)_2$ with potassium permanganate in acetic acid. This reaction adds a chemical step and may also be a safety hazard. For these reasons, this approach was not used. Alcohol **5** was next allowed to react with thionyl chloride to give the corresponding chloride, which, without isolation, was substituted with the anion produced from sodium hydride and dimethyl phosphite to give the desired phosphonate **6**. Sodium dimethyl phosphonate is an environmentally friendlier reagent than trimethyl phosphite, which stinks and produces carcinogenic chloromethane as a by-product in this (Arbuzov) reaction; the new conditions produce sodium chloride as the only by-product. The heterocycle was thus ready to be coupled with the side chain chiron.

6.2. FROM D-GLUCOSE

6.2.1. Medicinal Chemist's Synthesis

The side chain of these HMG-CoA reductase inhibitors resembles a carbohydrate, so it is not surprising that the medicinal chemists used D-glucose as a starting material to produce a chiron for the synthesis of the enantiopure side chain (Fig. 6.4). Following the procedure by Corey and co-workers (4) and Yang and Falck (5), D-glucose **7** was converted to D-glucal **8** (6,7). This reaction eliminated the unneeded hydroxyl group in the 2-position. For reasons we will understand in a moment, an axial methoxyl group was needed in the 1-position; to this end, mercuric acetate and sodium methoxide were added to the double bond of D-glucal in a *trans*-diaxial fashion to give **9**. Mercury was reduced with sodium borohydride, and the primary alcohol was protected with a trityl group to give **10**. Such a bulky protecting group was needed to carry out the next reaction selectively at the 3-hydroxyl group and not at the 4-hydroxyl group (which is now shielded by the bulky protecting group on the 6-hydroxyl).

1-(2,4,6-Triisopropylbenzenesulfonyl)imidazole was prepared from the corresponding sulfonic acid chloride in a separate step and added to **10**. This activating group was also chosen to be sterically bulky to improve the regioselectivity of the reaction and to direct it to the 3-hydroxyl group. Once the sulfonate was on the 3-position, the 4-hydroxyl group displaced it (with inversion) and formed epoxide **11**. The important stereochemical feature of this epoxide is that it is *syn* to the axial methoxyl group, a relationship that becomes important in the next step. The epoxide is opened and reduced *regioselectively* by lithium aluminum hydride from the 4-position.

Lithium, as a Lewis acid, apparently forms a 6-membered ring chelate with the axial methoxyl group and the epoxide, directing the hydride nucleophile to the 4-position. This selectivity explains why we needed the axial methoxyl group. The hypothesis is confirmed by the observation that the reaction is regioselective in diethyl ether but nonselective in tetrahydrofuran, which is a better coordinating agent and disrupts the intramolecular chelation of lithium. In this clever fashion, compound **12** was formed, containing all the stereochemical requirements of our chiron: the *syn*-3,5-diol and no functionality on carbons 2 and 4.

The rest of the steps were only functional group transformations: The 3-hydroxyl group was protected with another bulky protecting group, *t*-butyldiphenylchlorosilane, and the trityl-protecting group on the 6-hydroxyl was removed with trifluoroacetic acid to give **13**. The 6-hydroxyl group was oxidized with chromium(VI) oxide and pyridine to give **14**. This aldehyde was coupled in a Wittig reaction with the aromatic portion prepared previously (8).

Historically, the first such coupling was performed with a naphthalene analog (Fig. 6.5) (9). The aldehyde **14** was coupled with the anion of triphenylphosphonium **15** to give **16**, which was hydrolyzed with water and acetic acid, then oxidized with pyridinium chlorochromate to give lactone **17**. Deprotection of the alcohol with tetra-*n*-butylammonium fluoride and acetic acid gave **18**, and hydrolysis with ethanolic sodium hydroxide gave the drug substance **19**.

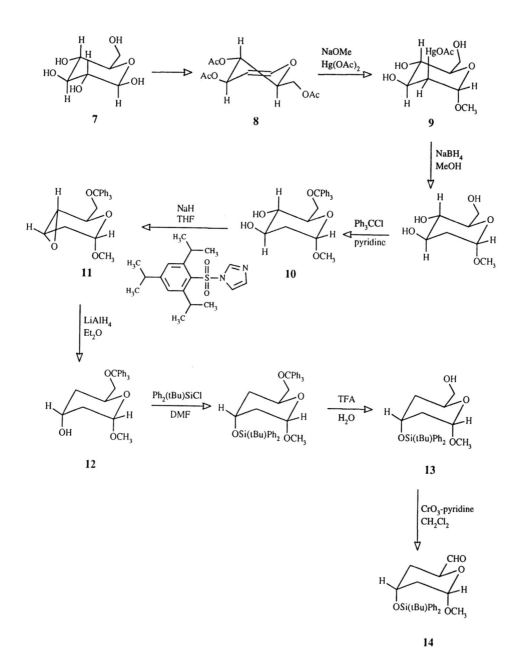

Fig. 6.4. Synthesis of side chain chiron from D-glucose.

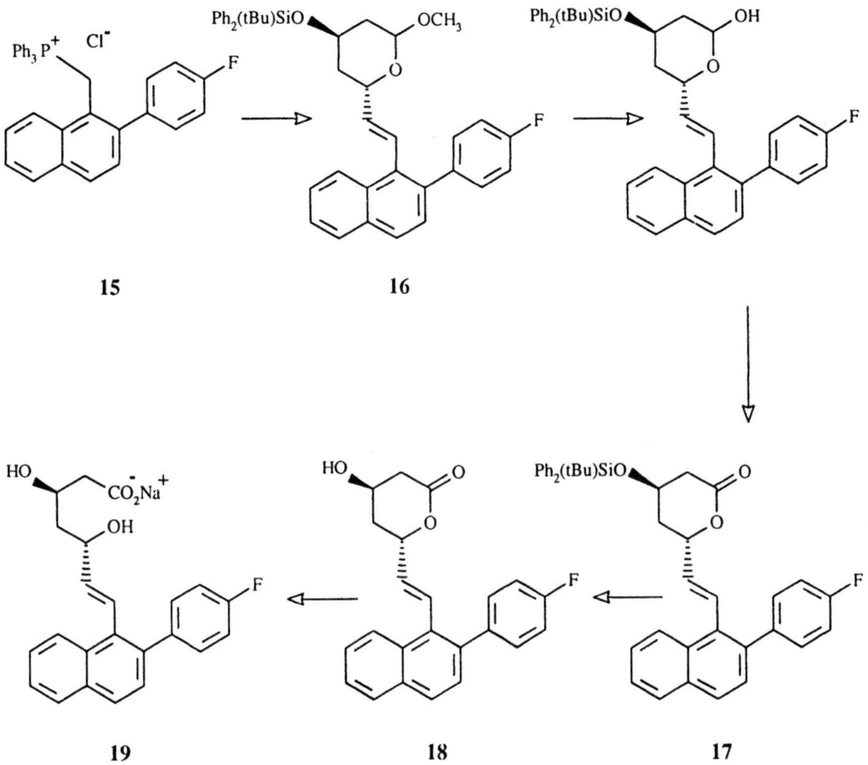

Fig. 6.5. Wittig coupling of the chiron.

However, among other problems of this synthesis that we will review in a moment, chiron **14** gave a mixture of isomers in both the Wittig coupling step (*E/Z* olefins **16**) and in the subsequent deprotection step (*cis/trans* lactones **18**). These mixtures of diastereomers had to be separated by a difficult chromatography. The mechanism of the latter isomerization is the reversible solvolytic opening/closing of the lactone in the allylic 5-position (Fig. 6.6).

This synthesis was anything but an ideal process. The mercuric acetate reagent may seem familiar from the synthesis used as the example in Table I.1: It would cost $48,135 to prepare 1 kg drug substance by this synthesis. It was too long to be economical, it used unacceptable reagents (mercuric acetate) and solvents (diethyl ether), and far too much of them at that. The sulfonyl reagent is expensive, the triphenylmethanol protecting group requires chromatography for its separation after removal, and so forth.

6.2.2. Developed Synthesis

Since in the early stages of drug development, time is most critical, and a new synthesis was not yet available (but see below), the discovery synthesis was developed

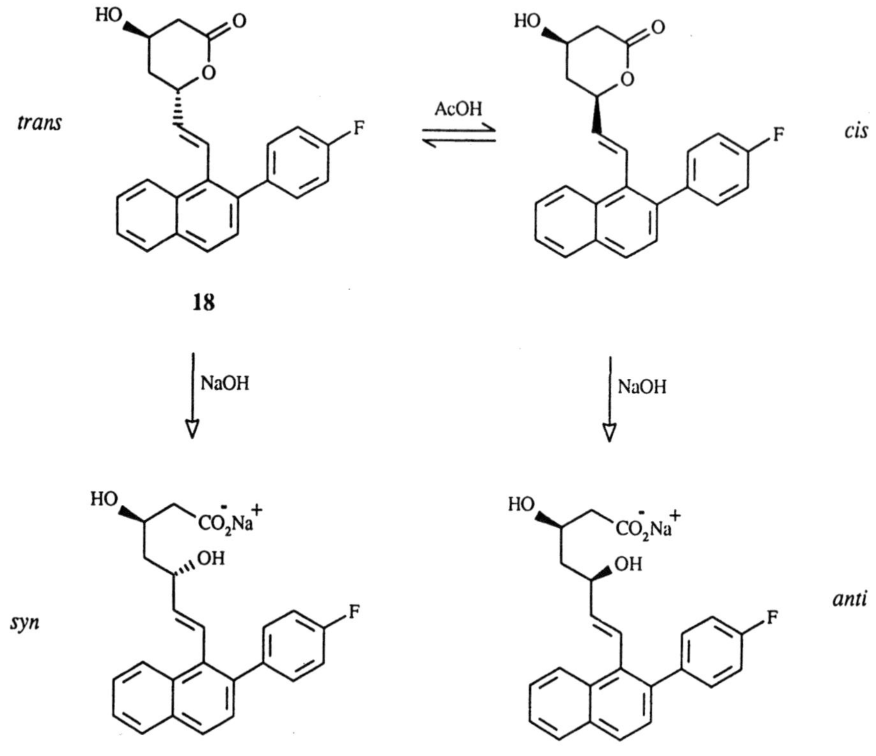

Fig. 6.6. Isomerization of the lactone.

as much as possible to enable us to produce a few kilograms of this valuable chiron, needed for several projects.

We first solved the most critical problem, the isomerizations that plagued the last few steps. We knew from the racemic synthesis that the *acyclic* synthon (see Fig. 3.7) was entirely stereoselective throughout the synthesis, so we converted the cyclic **14** to the acyclic chiron **22** (Fig. 6.7). First, the aldehyde was protected as an olefin (**20**), which was, as before, hydrolyzed and oxidized to lactone **21**; it was opened with methanol, the second carbinol was protected with *t*-butyldiphenylchlorosilane, and the aldehyde was restored by ozonolysis to give chiron **22**. This pathway was not chosen for scale up, again because considerable amounts of the *anti*-isomer were observed in the product. Of course, we have again formed an allylic lactone that isomerized, probably during the methanolysis.

Consequently, the synthesis had to be redesigned. The aldehyde was not introduced until the last step, and the 6-position was kept protected as a benzyl ether (Fig. 6.8). Thus the 6-hydroxyl group in **13** was protected as a benzyl ether, the acetal was hydrolyzed and oxidized to give lactone **23**, which was then opened by methanolysis (no isomerization); the resulting 5-hydroxyl was protected (**24**) and the 6-hydroxyl was deprotected (**25**) and oxidized to give the desired chiron **22**.

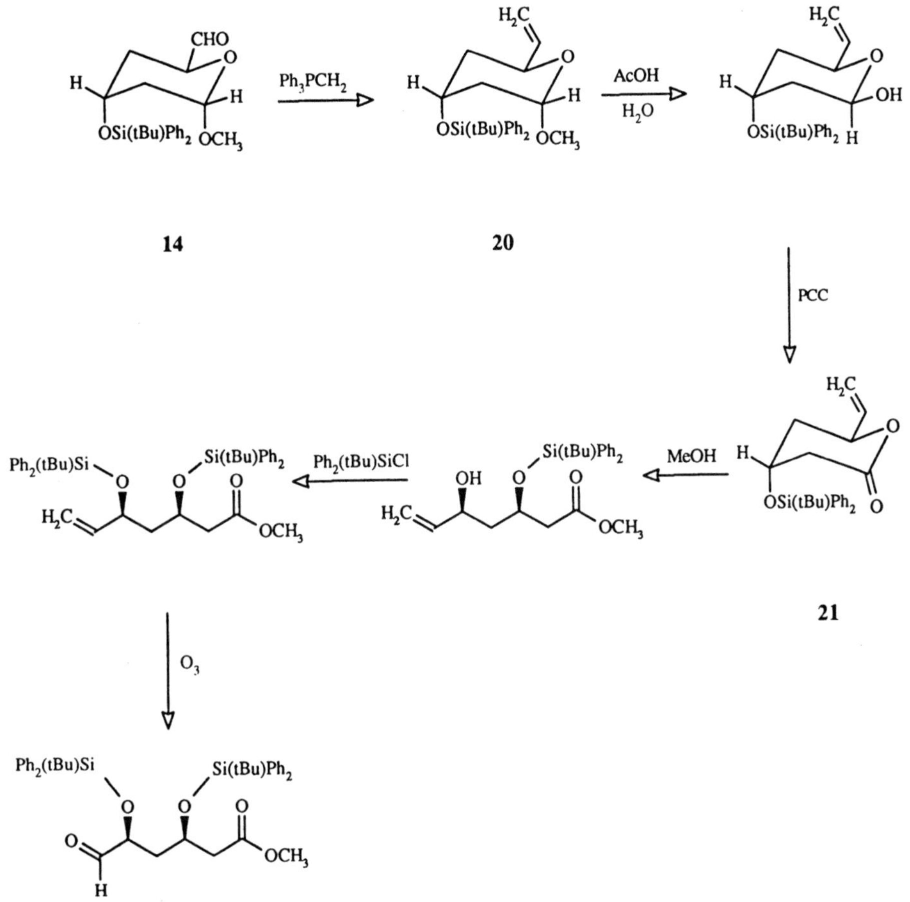

Fig. 6.7. Conversion of cyclic chiron to acyclic chiron (route I).

The remainder of the synthesis was analogous to the racemic synthesis (see Fig. 3.16), except that all of the subsequent intermediates and product were now enantiopure (10).

Other improvements were made throughout the synthesis. The conversion of D-glucose to D-glucal (Fig. 6.4) was much improved by replacing most of the inorganic reagents with HBr. The use of the mercury reagent in step 2 could not be avoided as other methods did not introduce the methoxyl axially. In the regioselective opening of the epoxide, diethyl ether offered the highest regioselectivity, as already noted; however, since it is an unsafe solvent, a substitute had to be found. *t*-Butyl methyl ether was the best compromise between reactivity, regioselectivity and safety.

The objective of a stereoselective synthesis was thus achieved, and process redesign and development described in this section reduced the cost of the synthesis

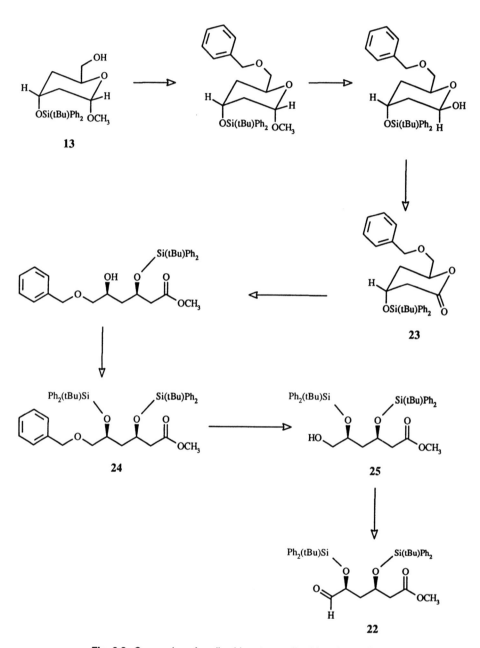

Fig. 6.8. Conversion of cyclic chiron to acyclic chiron (route II).

threefold. However, the synthesis was still too expensive and, at 25 steps, too long; the redesigned synthesis was shorter in the main sequence but longer for the side chain, so the total length remained unchanged. It was clearly not efficient: of the four stereogenic hydroxyls in D-glucose, we eliminated two (hydroxyls 2 and 4), inverted one (hydroxyl 3), and preserved only one (hydroxyl 5) in its natural configuration.

6.3. FROM L-MALIC ACID

Another natural product from which the 6-carbon-atom side-chain chiron can be derived is L-malic acid (11). It already contains an (S)-carbinol, to which all we need add is another syn-hydroxyl group, two carbon atoms (e.g., an acetate), and to adjust the oxidation state at one end of the molecule.

To this end, L-malic acid **26** was esterified with acetic acid chloride in methanol to give malate **27** (Fig. 6.9), and the proximal ester was reduced selectively with sodium borohydride, with chelation control by borane–dimethyl sulfide, to offer diol **28**. The primary hydroxyl was selectively protected with the triphenylmethyl (trityl) protecting group, to produce **29**. The chain in **29** was next extended (Fig. 6.10) by two carbon atoms in a condensation with *t*-butyl acetate, to give the hydroxy-keto-ester **30**. The *t*-butyl was selected because, as opposed to the methyl or ethyl esters, it did not lactonize and, therefore, did not epimerize as readily in the penultimate step of the synthesis (during de-silylation). Intermediate **30** was diastereoselectively reduced with sodium borohydride, with chelation control by diethylmethoxyboron, in tetrahydrofuran–methanol (4:1) at $-76°C$, to obtain diol **31**

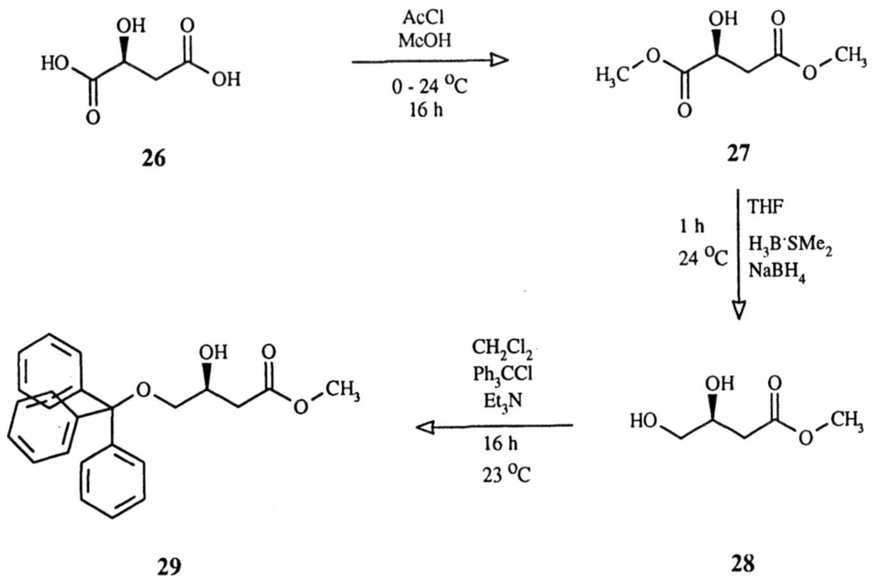

Fig. 6.9. Synthesis of chiron from L-malic acid (part I).

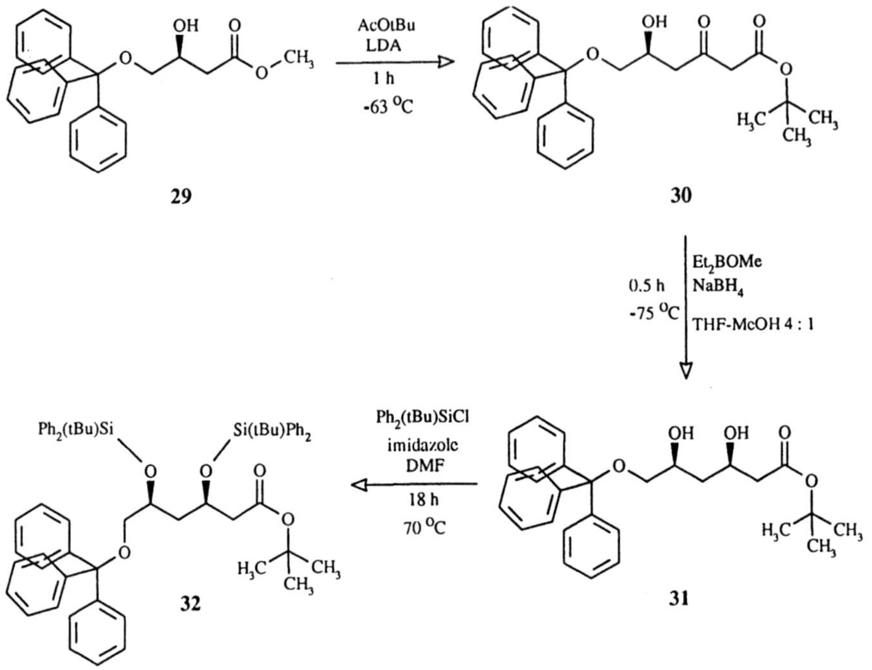

Fig. 6.10. Synthesis of chiron from L-malic acid (part II).

of 99% enantiomeric and diastereomeric purity. This diastereoselective reaction had been invented and already optimized for our previous racemic synthesis (see Tables 3.11 to 3.15). The diol was again protected as *t*-butyldiphenylsilyl ethers **32**.

The trityl group was removed with trifluoroacetic acid (Fig. 6.11), and the primary alcohol **33** was oxidized with pyridinium chlorochromate to give the desired chiron **34**, which is the *t*-butyl analog of **22**.

We were at last ready to attach the side chain to the heterocycle **6** via a Horner–Emmons reaction (Fig. 6.12) (12). The advantages of these conditions over the Wittig coupling have already been discussed in the racemic synthesis in Chapter 3; one of the advantages was the higher *E*-diastereoselectivity. The advantage of the bulky *t*-butyldiphenylsilyl protecting group in **34** has also already been discussed: It prevents hydration of the aldehyde, which occurs rapidly with α-hydroxylaldehydes. This effect of the bulky protecting group must be hydrophobic and not steric since the Horner–Emmons reaction proceeds with high yield (78%) and rate (0.2 h at −5°C) although both components **34** and **6** are sterically crowded. Thus an anion of **6** was formed with lithium diisopropylamide (by reverse addition to prevent methylation) which condensed with aldehyde **34** to form **35**. The silyl ethers were removed with tetra-*n*-butylammonium fluoride in acetic acid and acetonitrile. This reaction was slow (30 h at 64°C), but the *t*-butyl ester ensured that no lactone formed and that no isomerization to the *anti*-isomer occurred. Diol **36** was thus obtained in 82% yield and with 99% enantiomeric and diastereomeric purity.

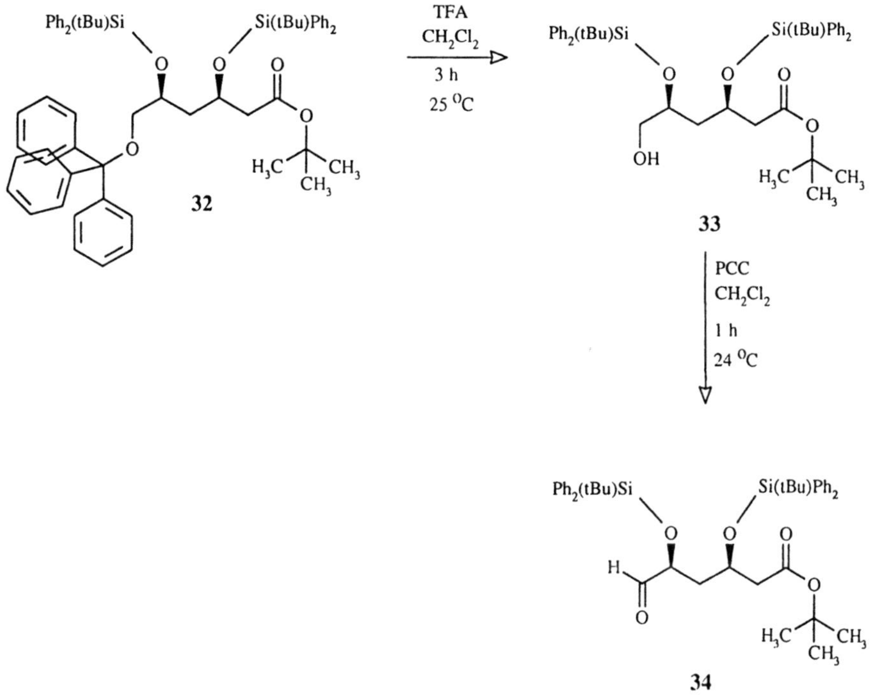

Fig. 6.11. Synthesis of chiron from L-malic acid (end).

It was saponified with methanolic sodium hydroxide to give the desired drug substance **37**. Though it is a *t*-butyl ester, the saponification occurs unusually fast (1 h at 25°C). The reason for this fast reaction is the general acid catalysis provided by the β- and the δ-hydroxyl groups; we have shown that removal of one or both of the hydroxyl groups makes the saponification **36** → **37** progressively slower.

The new synthesis of **37** from L-malic acid was quite impressive: The total number of steps was only 17 (since the side-chain chiron **34** was made in only 8 steps), and the cost had been reduced by another factor of five. But everything is relative: Both the length and the cost were still too high.

6.4. ASYMMETRIC SYNTHESES

Another (biomimetic) strategy for constructing the seven-carbon side chain in drug substance **37** is shown in Figure 6.13. Starting with an aldehyde on the imidazole, three aldol condensations with acetate would produce the desired precursor **36**. If the second of these reactions was to be asymmetric, **36** would be made enantiomerically pure.

Accordingly, intermediate **4** was condensed with ethyl acetate to give **38**, reduced to allylic alcohol **39**, and oxidized to the conjugated aldehyde **40** (Fig.

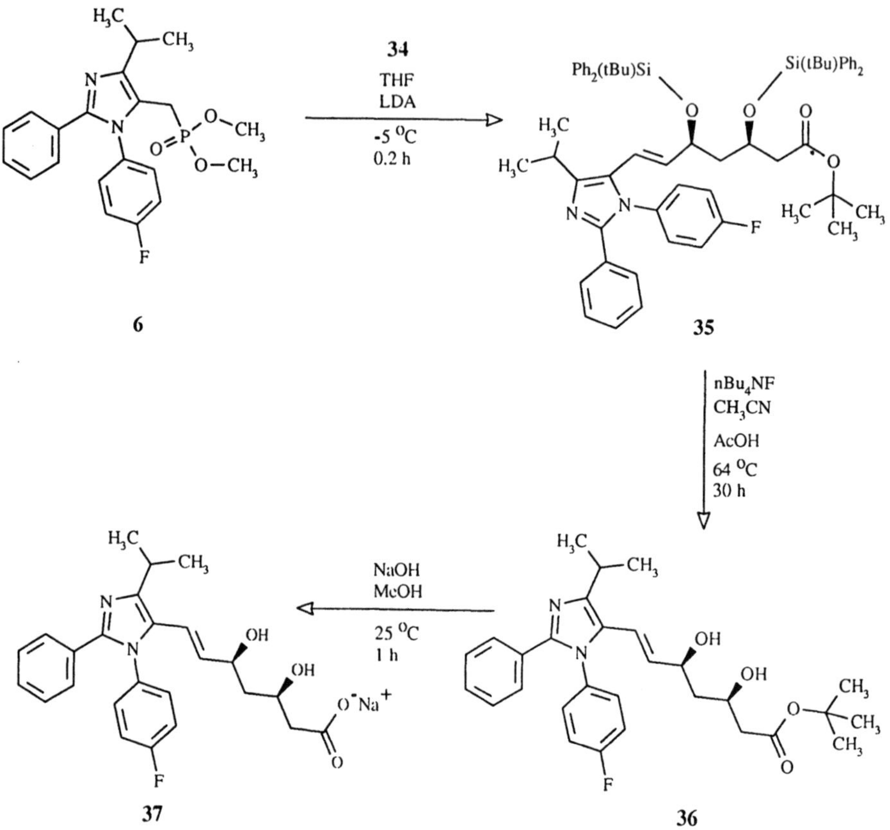

Fig. 6.12. Horner–Emmons coupling toward drug substance.

6.14). To generate the stereogenic carbinol, an asymmetric aldol was needed next. At the time we started this research, few asymmetric aldol reactions with acetate (i.e., without an α-substituent) were known; the best were by Evans (13) and by Mukaiyama (14,15) (Fig. 6.15).

Unfortunately, the enantiomeric selectivity of the Mukaiyama aldol was not sufficiently high with our aldehyde. For practical and economic reasons (the requirement was no chromatography and high yields), our demanding goal was an enantiomeric excess of >90%.

Another, at the time already famous, asymmetric reaction was attempted, the Sharpless epoxidation (Fig. 6.16) (16). The first stereogenic carbinol **41** was formed with excellent (92%) enantiomeric excess (17); however, the second (diastereoselective aldol reaction **42 → 43**) was totally nonselective under modified Reformatsky conditions, and a 1 : 1 mixture of *syn*-**43** and *anti*-**43** was formed.

This synthesis illustrates and supports Sharpless's philosophy that diastereoselectivity is more difficult to achieve than is enantioselectivity (18). In other words,

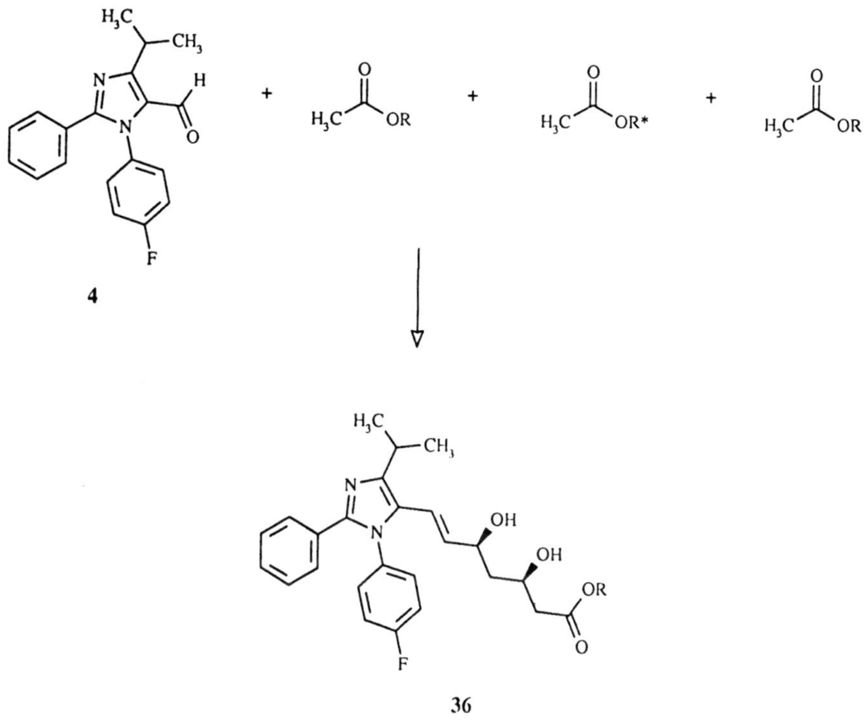

Fig. 6.13. Asymmetric synthesis of HMG-CoA-reductase inhibitors.

one should use an asymmetric reagent to establish stereochemistry and not rely only on the asymmetry in the molecule, as the Woodwardian school did. It also pointed to the direction for our research; namely, we needed to develop an asymmetric aldol reaction. Fortunately, several asymmetric acetate aldol methodologies had been invented in Europe in the 1980s (19–23). With our molecules the best results were obtained with Braun's (19) asymmetric aldol reaction, which uses (S)-2-acetoxy-1,1,2-triphenylethanol as the asymmetric reagent, prepared from L-mandelic acid.

6.5. FROM L-MANDELIC ACID

L-Mandelic acid **44** was esterified to **45** (Fig. 6.17) and allowed to react with two equivalents of phenyl Grignard to yield triphenylethylene glycol **46**. Acetylation with acetyl chloride produced Braun's reagent **47**.

The reaction between conjugated aldehyde **40** and the lithium anion of **47** made the S-hydroxyl ester **48** in an excellent enantiomeric excess of 96% (i.e., a 98:2 ratio of enantiomers) (Fig. 6.18). Ester **48** was converted to the methyl ester **49** with methanol and potassium carbonate and was crystallized until enantiomerically pure, as determined by shift-reagent NMR. By adding acetic anhydride to the mother

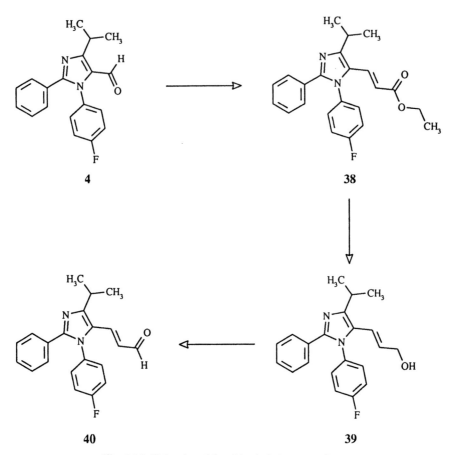

Fig. 6.14. Extension of the side chain by two carbons.

liquors, the insoluble **47** was recovered as a solid in 80% yield without loss of enantiomeric purity and could be reused, as required for a good asymmetric synthesis. Empirically trying out both the (*S*)- and (*R*)-mandelates, we proved that reagent **47**, derived from (*S*)-mandelate, gave the desired enantiomer of **49**.

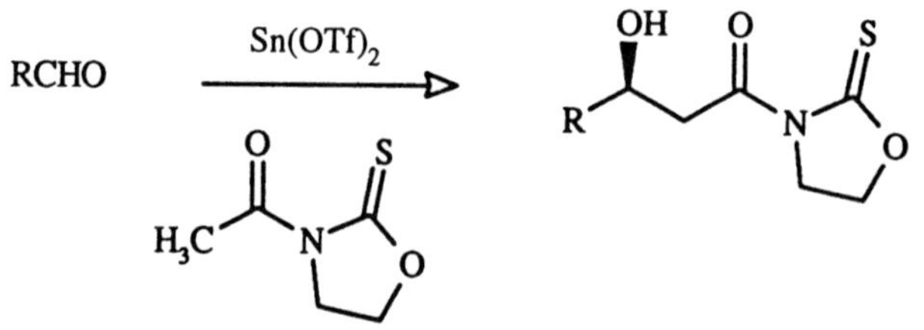

Fig. 6.15. Mukaiyama's asymmetric aldol reaction.

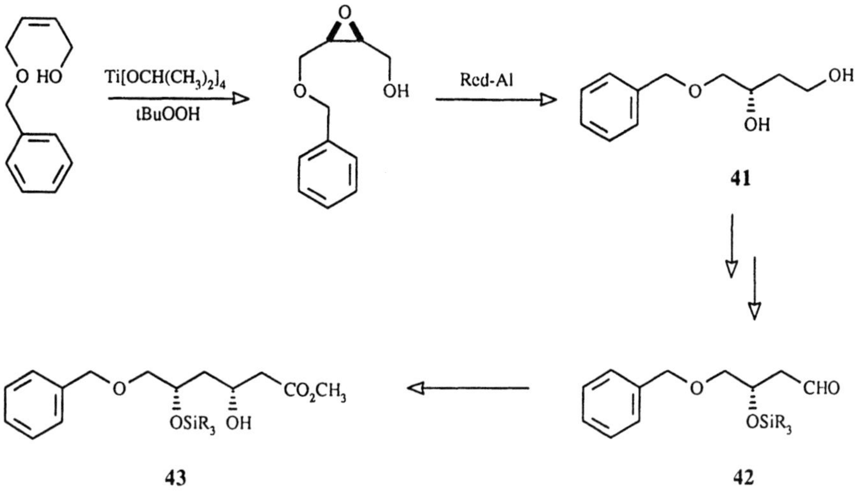

Fig. 6.16. Sharpless's asymmetric epoxidation approach.

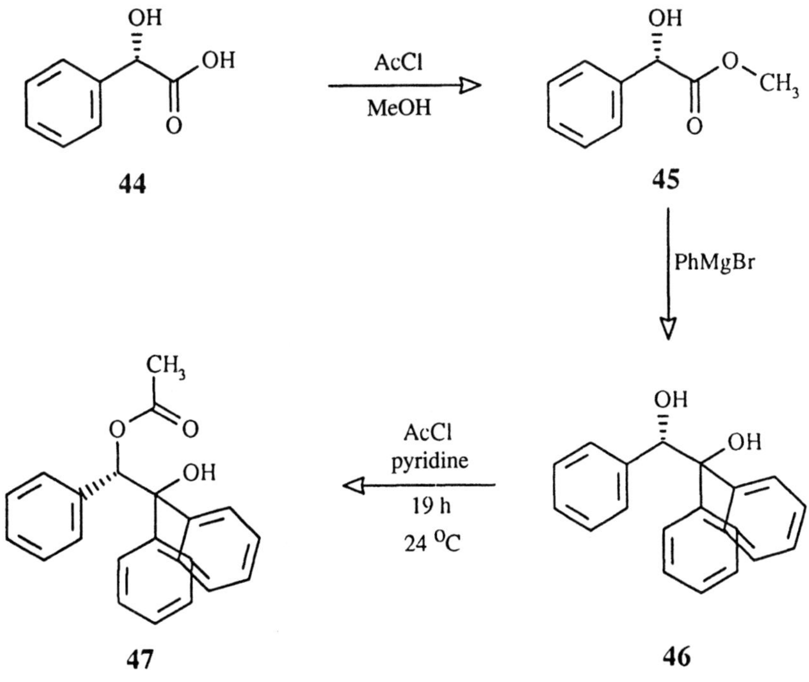

Fig. 6.17. Synthesis of Braun's reagent.

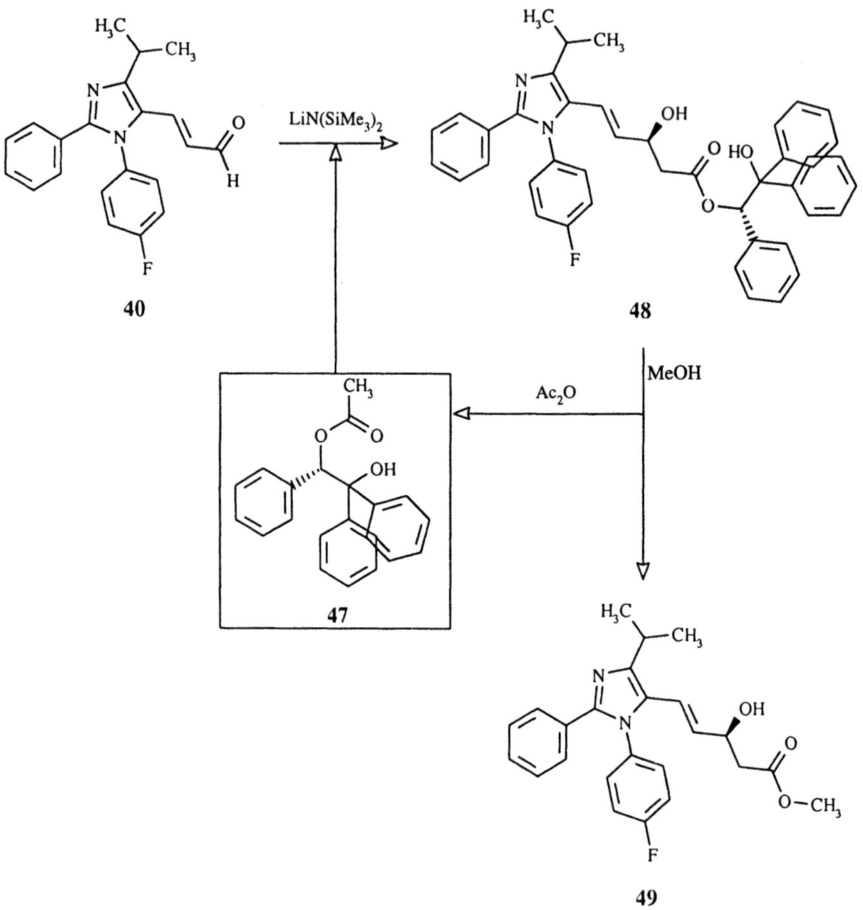

Fig. 6.18. Braun's asymmetric aldol reaction.

We optimized Braun's asymmetric aldol reaction (**40 → 48**) for our aldehyde by varying the base, the amount of base, and the reaction temperature (24). Not surprisingly, the lowest temperature (−100°C) gave the best enantioselectivity (98:2), and the best base was lithium hexamethyldisilazide (Table 6.1), prepared from n-butyllithium and the amine. Less obvious was our observation that an excess of n-butyllithium or the addition of LiCl was detrimental to the enantioselectivity; apparently these reagents disrupt the lithium chelate of asymmetric reagent **47** that is critical for high enantioselectivity. A slight excess of the amine over n-butyllithium is, therefore, prescribed. As no suitable crystals of this lithium chelate could be obtained so far, the exact structure of Braun's reagent is not known.

We also tried to optimize the reaction by varying the substituents on the reagent, but with no success. All three phenyls are needed for the high stereoselectivity (Table 6.2).

Table 6.1. Optimization of Braun's Asymmetric Aldol Reaction 40 → 48

Amine, equiv	n-BuLi, equiv	Additive, equiv	Temperature, °C	ee, %
Dicyclohexylamine (3)	3	—	−78	84
2,2,6,6-Tetramethylpiperidine (3)	3	—	−78	90
Diisopropylamine (3)	3	—	−78	92
Hexamethyldisilazane (3)	3	—	−78	94
Hexamethyldisilazane (4)	3.6	—	−100	96
Hexamethyldisilazane (3.6)	7.2	—	−100	70
Hexamethyldisilazane (4)	3.6	LiCl	−100	70

The synthesis concluded with the third condensation (Claisen, this time) of t-butyl acetate with intermediate **49** to produce the β-hydroxy-ketone **50** (Fig. 6.19). It was reduced using our original syn-selective reduction (NaBH$_4$) by chelation (Et$_2$BOMe) to make diol **36**. The advantages of the t-butyl ester were itemized for the racemic synthesis in Chapter 3. It offers a higher stereoselectivity in the reaction **50 → 36** and makes **36** crystalline, so that it can be purified to a diastereomeric excess of 98%. Finally, a saponification of **36** with methanolic sodium hydroxide gave drug substance **37**, containing less than 0.5% of the distomer and less than 0.5% of the anti-diastereomer.

To determine the stereochemical purity of **36**, all four of the stereoisomers were synthesized for analytical reference. All four were prepared (11,25) from the same intermediate **50** by alternating Prasad's (26) syn-selective reduction and Evans's (27) anti-selective reduction of β-hydroxy ketones (Fig. 6.20).

Thus **50** was reduced by Prasad's chelation-control reaction (Et$_2$BOMe, NaBH$_4$) to the syn-(S,R)-isomer **36**. It was in turn oxidized at the allylic position with 2,3-dichloro-5,6-dicyano-1,4-benzoquinone (DDQ), and the resulting reversed hydroxy ketone was reduced by Evans's conditions (Me$_4$N$^+$HB$^-$(OAc)$_3$) to give the anti-(R,R)-isomer **51**. Alternatively, **50** was first reduced by Evans's reagent to give the anti-(S,S)-isomer **52**, which was oxidized with DDQ, and the allylic ketone was reduced by Prasad's reaction to give the syn-(R,S)-isomer **53**.

Table 6.2. Optimization of Braun's Reagent 47 in Reaction 40 → 48

Vicinal Dialkyl	Stereogenic Alkyl	ee, %
Methyl	Phenyl	60
Isopropyl	Phenyl	58
Phenyl	Methyl	58
Phenyl	Phenyl	96
Phenyl	o-Toluyl	74
Phenyl	m-Toluyl	60
Phenyl	p-Toluyl	88
Phenyl	p-Methoxyphenyl	96

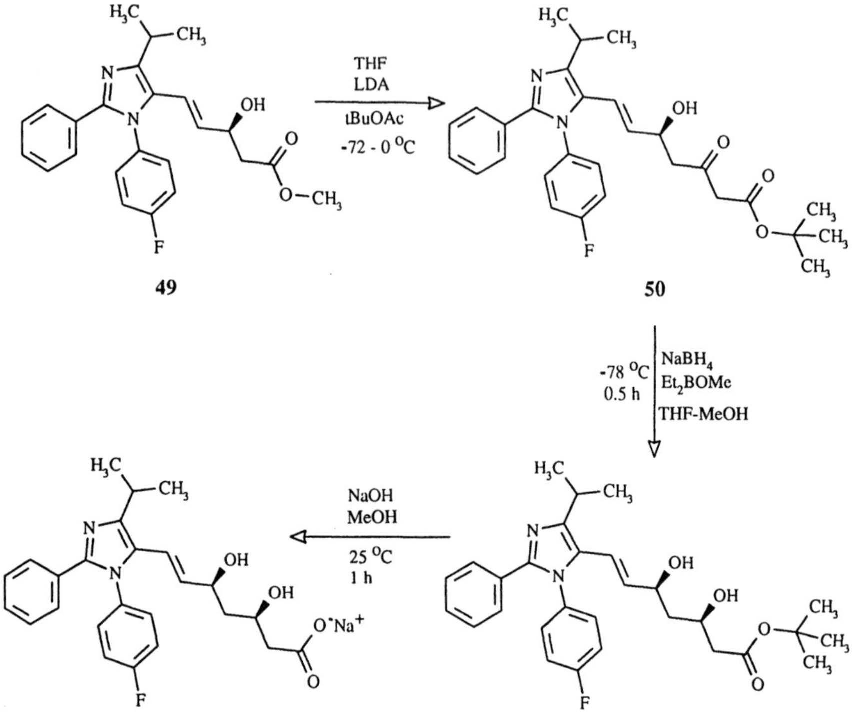

Fig. 6.19. Asymmetric synthesis of HMG-CoA-reductase inhibitor (grand finale).

6.6. CONCLUSIONS

The moral of this chapter is that asymmetric syntheses are usually more efficient than chiral pool strategies. An asymmetric synthesis can be as short as a racemic synthesis (just replace one reagent with an asymmetric reagent), whereas the 25-step synthesis starting from D-glucose was most inefficient: we used only one stereogenic hydroxyl of glucose, the others were all either inverted or discarded.

The other lesson is that reducing the number of steps in a synthesis—by designing a shorter synthesis—is probably the most effective method for reducing the cost of a process (Table 6.3).

We saw in Table I.1 that the original synthesis from discovery was expensive and that it contained many unacceptable ingredients. The usual process development (maximizing yields, minimizing the amounts of reagents and solvents, and slight modifications in the synthetic pathway) reduced the cost of raw materials by a factor of only about 3, from \$48,135 to \$16,845 (Table 6.3). A new, shorter synthesis, on the other hand, can bring the price down by a factor of 5, from \$16,845 to \$3,505 or, if the synthesis is shortened further, by a factor of 12, to \$1,436. Not by coincidence, the latter synthesis—from L-mandelic acid—is an asymmetric synthesis that uses a recyclable reagent.

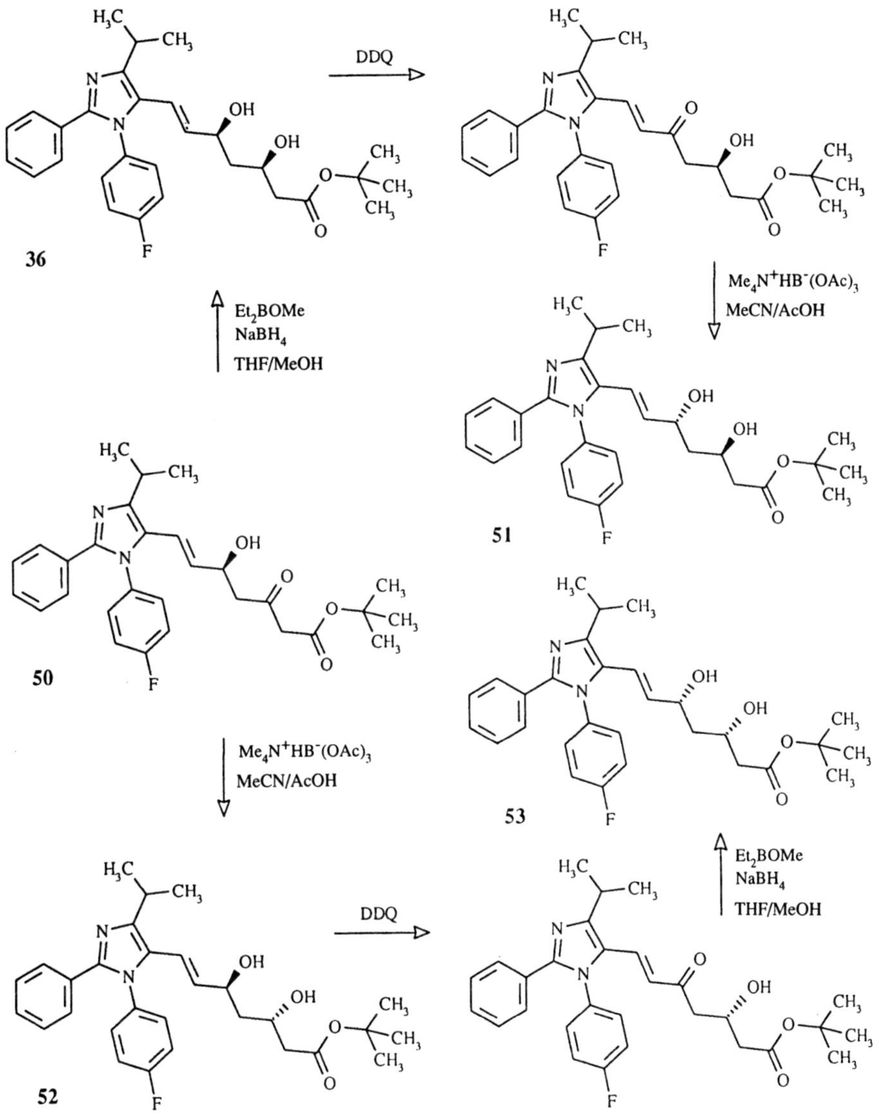

Fig. 6.20. Synthesis of all four isomers.

Table 6.3. Process Research and Development Reduces the Cost of a Synthesis

Stage of Process	Number of Steps	Cost of Chemicals	Starting Material	Synthetic Strategy
Discovery	15 + 10 = 25	$48,135	D-Glucose	Chiral pool
Development	9 + 16 = 25	$16,845	D-Glucose	Chiral pool
New Synthesis	9 + 8 = 17	$3,505	L-Malic Acid	Chiral pool
New Synthesis	11	$1,436	L-Mandelic Acid	Asymmetric synthesis

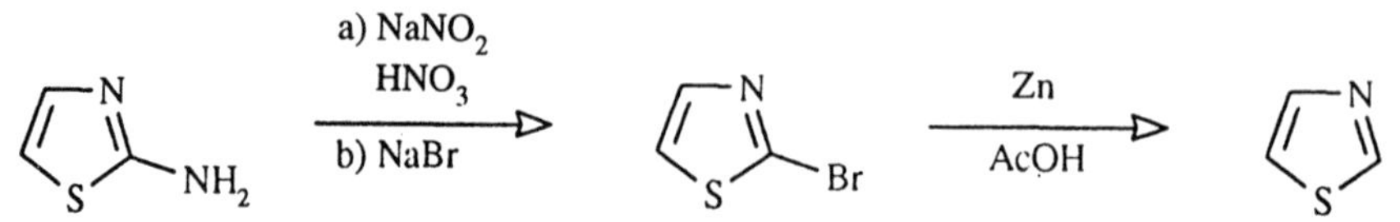

Fig. 6.21. Synthesis of thiazole.

Attentive readers may remember that, in Chapter 1, I suggested that a good strategy for lowering the cost of a synthesis was to make it convergent, but the first three syntheses in this chapter (summarized in Table 6.3) are convergent yet the most expensive. Well, convergent syntheses are more economical than linear syntheses if the total number of steps is *the same*. The shortest synthesis will usually be the cheapest, even if it is linear, as in the present examples.

One obvious exception to this rule of thumb is a "backward integration": if a starting material is too expensive, then *adding* steps to synthesize it in-house can often lower the cost of a process. For example, the commercial price of unsubstituted thiazole is about $5,000/kg, of 2-bromothiazole about $500/kg, and of 2-aminothiazole about $25/kg (Fig. 6.21). Adding a couple of steps to the synthesis can thus *lower* the cost by a factor of 20 (after adding in the cost of labor).

This chapter illustrates the great power of process development in lowering costs and the great value added by chemical research and development.

BIBLIOGRAPHY

1. F. G. Kathawala, "HMG-CoA Reductase Inhibitors: An Exciting Development in the Treatment of Hyperlipoproteinemia," *Med. Res. Rev.*, **11**, 121–146 (1991).

2. S. Hanessian, *Total Synthesis of Natural Products: The 'Chiron' Approach*, Pergamon Press, Oxford, UK, 1983.

3. World Pat. WO 8607054 (1986), "New Imidazole Analogues of Mevalonolactone, Useful for Treating Hyperlipoproteinemia and Atherosclerosis," J. R. Wareing (to Sandoz).

4. E. J. Corey, L. O. Weigel, A. R. Chamberlin, and B. Lipshutz, "Total Synthesis of (−)-N-Methylmaysenine," *J. Am. Chem. Soc.*, **102**, 1439–1441 (1980).

5. Y.-L. Yang and J. R. Falck, "Mevinic Acids and Analogs: Preparation of a Key Chiral Intermediate," *Tetrahedron Lett.*, **23**, 4305–4308 (1982).

6. U.S. Pat. 4,474,971 (Oct. 2, 1984), "New 2-Formyl-4,6-diprotected-hydroxytetrahydropyran Compounds—Useful as Intermediates for Hypocholesterolemic 4-Hydroxypyranone Derivatives," J. R. Wareing (to Sandoz).

7. U.S. Pat. 4,625,039 (Nov. 25, 1986), "New 6-Oxotetrahydropyran-2-yl Aldehyde Compounds Are Intermediates for Antihypercholesterolemic 6-*trans*-Olefinically Substituted Tetrahydropyran-2-one Compounds," C. F. Jewell and J. R. Wareing (to Sandoz).

8. U.S. Pat. 4,755,606 (July 5, 1988), "New Imidazolyl-3,5-di(diphenylbutylsilyloxy) carboxylate Esters, Useful as Intermediates for Cholesterol Biosynthesis Inhibitors," J. R. Wareing (to Sandoz).

9. Ger. Pat. DE 3,525,256 (1986), "New Naphthyl Analogues of Mevalonolactone, Useful as Cholesterol Biosynthesis Inhibitors," P. L. Anderson (to Sandoz).

10. Eur. Pat. 244,364 (1987), "Production of 7-Substituted 3,5-Dihydroxy-6-heptene-1-oic Acid Derivatives, Including New Optically Pure Isomers, from Protected 6-Oxo-3,5-dihydroxyhexanoate Esters," K.-M. Chen, G. E. Hardtmann, P. K. Kapa, G. T. Lee, J. Linder, and S. Wattanasin (to Sandoz).

11. K. Prasad, K.-M. Chen, O. Repič, and G. E. Hardtmann, "A Highly Stereoselective Route to the Four Stereoisomers of a Six-Carbon Synthon," *Tetrahedron Asym.,* **1,** 307–310 (1990).

12. G. T. Lee, J. Linder, K.-M. Chen, K. Prasad, O. Repič, and G. E. Hardtmann, "A General Method for the Synthesis of *syn-(E)*-3,5-Dihydroxy-6-heptenoates," *SynLett,* **1990,** 508.

13. D. A. Evans, J. Bartroli, and T. L. Shih, "Enantioselective Aldol Condensations. 2. *Erythro*-selective Chiral Aldol Condensations via Boron Enolates," *J. Am. Chem. Soc.,* **103,** 2127–2129 (1981).

14. T. Mukaiyama, "Metal Enolates in Organic Synthesis," *Pure Appl. Chem.,* **55,** 1749–1758 (1983).

15. R. W. Stevens and T. Mukaiyama, "Efficient Asymmetric Induction *via* Coordination of Chiral Diamine to Tin(II) Enolate: A Highly Enantioselective Synthesis of 2-Substituted Malates," *Chem. Lett.,* **1983,** 1799–1802.

16. R. M. Hanson and K. B. Sharpless, "Procedure for the Catalytic Asymmetric Epoxidation of Allylic Alcohols in the Presence of Molecular Sieves," *J. Org. Chem.,* **51,** 1922–1925 (1986).

17. K. Prasad and O. Repič, "Asymmetric Synthesis of (3*R-trans*) and (3*S-cis*)-Hydroxy-5-pentanolides," *Tetrahedron Lett.,* **25,** 3391–3394 (1984).

18. K. B. Sharpless, lecture given at the Nobel Symposium on Asymmetric Organic Synthesis in Karlskoga, Sweden, Sept. 2, 1984.

19. M. Braun and R. Devant, "The (*R*)- and (*S*)-2-Acetoxy-1,1,2-triphenylethanols—Effective Synthetic Equivalents to a Chiral Acetate Enolate," *Tetrahedron Lett.,* **25,** 5031–5034, (1984).

20. G. Helmchen, U. Leikauf, I. Taufer-Knöpfel, "Enantio- and *anti*-Diastereoselective Aldol Additions of Acetates and Propionates *via* O-Silylketene Acetals," *Angew. Chem.,* **97,** 874–876 (1985).

21. W. Oppolzer and J. Marco-Contelles, "Asymmetric and *anti*-Selective Aldolizations of Acetates and Propionates," *Helv. Chim. Acta,* **69,** 1699–1703 (1986).

22. S. G. Davies, I. M. Dordor, P. Warner, "Chiral Acetate Enolate Equivalent for the Synthesis of β-Hydroxy Acids," *J. Chem. Soc. Chem. Commun.,* **1984,** 956–957.

23. M. T. Reetz, A. Jung, "1,3-Asymmetric Induction in Addition Reactions of Chiral β-Alkoxy Aldehydes: Efficient Chelation Control *via* Lewis Acid Titanium Reagents," *J. Am. Chem. Soc.,* **105,** 4833–4835 (1983).

24. "Stereoselective Aldol Reaction with Chiral Acetates," K. Prasad, K.-M. Chen, O. Repič, and G. E. Hardtmann, *Tetrahedron Asym.,* **1,** 703–706 (1990).

25. K.-M. Chen and K. Prasad, personal communication, 1989.

26. K.-M. Chen, G. E. Hardtmann, K. Prasad, O. Repič, and M. J. Shapiro, "1,3-*syn*-Diastereoselective Reduction of β-Hydroxy Ketones Utilizing Alkoxydialkylboranes," *Tetrahedron Lett.,* **28,** 155–158 (1987).

27. D. A. Evans, K. T. Chapman, and E. M. Carreira, "Directed Reduction of β-Hydroxy Ketones Employing Tetramethylammonium Triacetoxyborohydride," *J. Am. Chem. Soc.,* **110,** 3560–3578 (1988).

7

RADIANT

7.1. BASICS

To study the fate and distribution of a drug in the body or in the environment, drug substances for these experiments are often labeled with isotopes. The most commonly used isotopes are stable isotopes such as deuterium (^2H) or carbon-13 (^{13}C); samples labeled with these isotopes and their metabolites can often be analyzed and quantified by mass spectroscopy. Alternatively, radioisotopes like tritium (^3H) and carbon-14 (^{14}C) can be used, as they can be detected at small levels by measuring their (β) radiation. Tritium has a maximum specific activity of 29.1 Ci/mmol and a half-life of 12.3 years, whereas ^{14}C has a specific activity of 62.4 mCi/mmol and a half-life of 5730 years. One can build into one molecule several radioisotopic atoms at once, but there is a practical limit to the specific activity of about 120 Ci/mmol, because of the instability of such molecules. Most often the labeled compound is highly diluted with the unlabeled substance; this is indicated by brackets, e.g., [^{14}C]carbon dioxide.

The underlying assumption in labeling experiments is that the isotopically labeled substance behaves exactly as the unlabeled substance. This hypothesis is not always true, especially if the isotope label is at or near the metabolized site in the molecule (secondary isotope effects).

Sometimes it is even possible to separate isotopomers by high-pressure liquid chromatography (HPLC) (1,2). Reverse-phase HPLC analysis of a [5-^3H]-labeled drug substance **1** (Fig. 7.1) revealed isotopic fractionation, that is, the labeled and unlabeled molecules had different retention times. This observation was validated by physically separating the labeled and unlabeled molecules by HPLC. It was found that this separation was pH dependent, which implies that the pK_a values of the labeled and unlabeled compound are different. [N-C$^3H_3$]-labeled **1** also exhibited

155

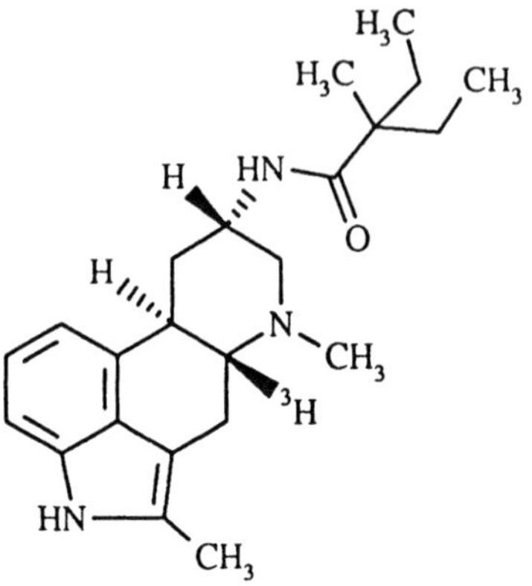

Fig. 7.1. Isotope effect on pK$_a$.   **1**

an isotope effect but to a different extent. Nevertheless, labeled compounds are usually useful in absorption, distribution, metabolism, and excretion (ADME) studies.

The radiosynthesis group will often align itself with chemical development. Not only do they share a common theme, organic synthesis, they also share unlabeled ("cold") synthetic intermediates. For this reason, radiosyntheses are often designed to use an intermediate that is common with the cold synthesis and that appears as late in the synthesis as possible, to minimize the number of radioactive steps.

Synthetically, the easiest strategy for introducing a label is to replace a hydrogen with deuterium or tritium; on the other hand, hydrogen labels are also lost most easily, as many metabolic reactions in the body remove hydrogen atoms from molecules in the form of water. It is, therefore, important to find a biologically stable position for the label; any loss of the label through this mechanism can lead to a false quantification of the drug in biological samples.

For this reason, carbon isotopes are often preferred. However, syntheses that incorporate carbon isotopes into the drug substance are usually longer as the carbon label, by definition, must be built into the carbon skeleton of the molecule.

The following examples will illustrate these principles.

7.2. PURCHASE LABELED REAGENTS

Often the same synthesis can be used for labeled as for the unlabeled drug substance by simply replacing one reagent with a commercially available isotopomer. For example, in the synthesis of SDZ 269-527, thiophosgene was replaced with [^{14}C]thiophosgene (Fig. 7.2) (3).

Butalbital was labeled by replacing urea, the reagent in one synthetic step, with [^{14}C]urea (Fig. 7.3) (4). Methyl iodide was replaced with ^{13}C^{2}H$_3$I in the synthesis of SDZ WAG 994 (Fig. 7.4) (5). Carnitine was replaced with (^{2}H$_9$)carnitine to prepare labeled compound **2** (Fig. 7.5) (6).

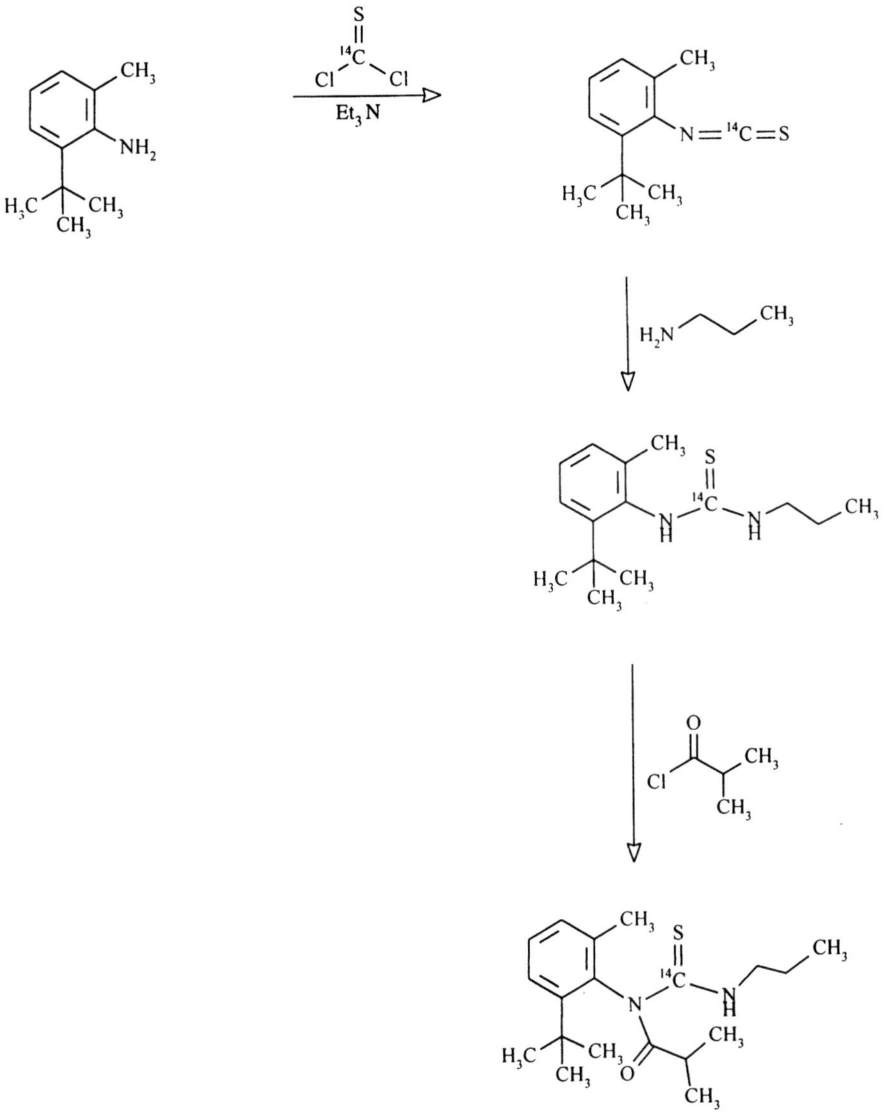

Fig. 7.2. Radiosynthesis with labeled thiophosgene.

The purchased labeled reagent sometimes needs a slight modification before it is inserted. For example, ($^{13}C_6$)cyclohexylamine was prepared from commercially available ($^{13}C_6$)nitrobenzene by hydrogenation, then used in the labeling of drug substance SDZ WAG 994 (Fig. 7.6) (7). ($^{13}C_2{}^2H_4$)Ethylene glycol was converted to the cyclic phosphate **3** to build a labeled choline phosphate into the phospholipid **4**, which was discussed in Chapter 5 (Fig. 7.7) (8). ($^2H_9$)t-Butyl chloride was converted to the corresponding Grignard reagent and added to tolunitrile (Fig. 7.8) (9).

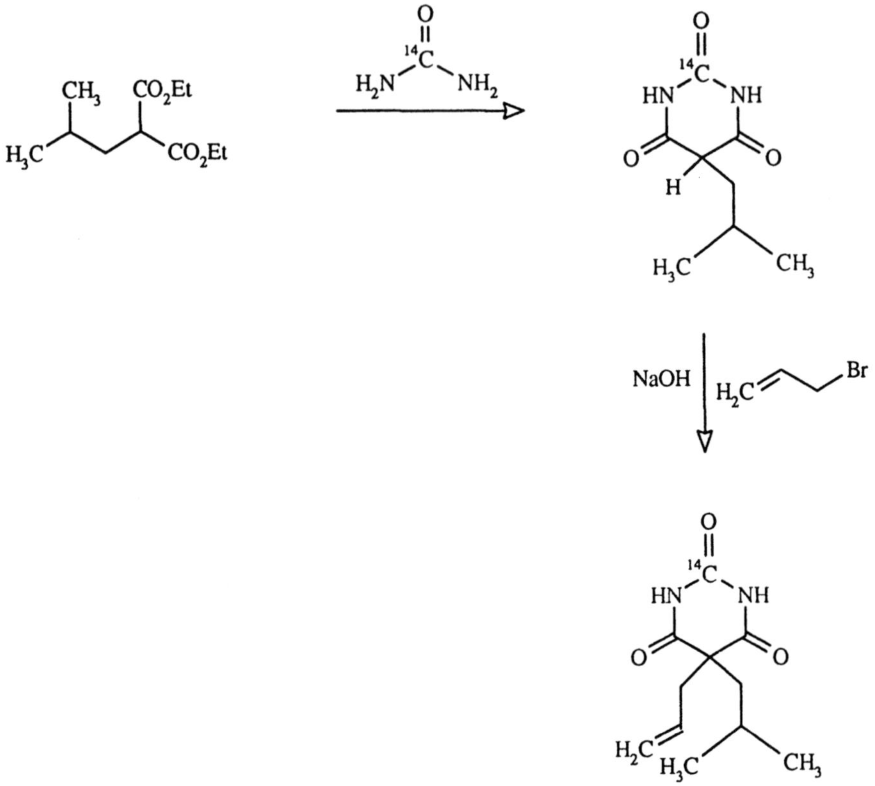

Fig. 7.3. Radiosynthesis of butalbital.

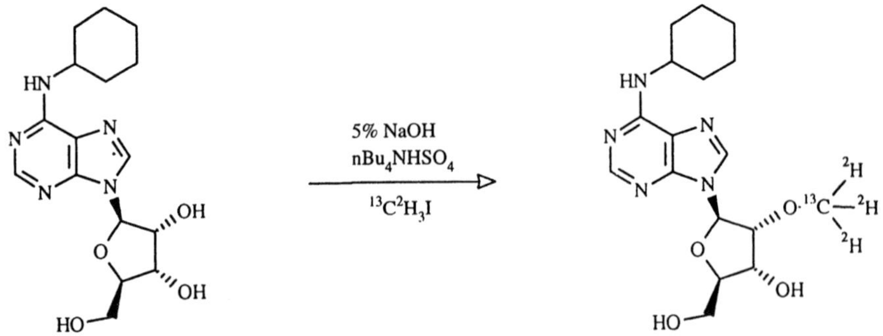

Fig. 7.4. Methylation with stable isotopes.

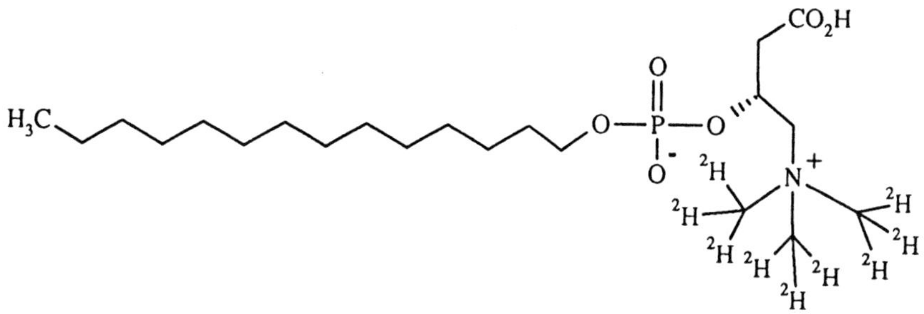

2

Fig. 7.5. Drug substance prepared from deuterated carnitine.

7.3. H-LABELS

Isotope labels ^{2}H or ^{3}H can be introduced most conveniently in an acid–base reaction, namely by deprotonation with a base and addition of an appropriately labeled acid. The method can be selective, e.g., by replacing the most acidic proton in the molecule (Figs. 7.9 and 7.10) (10).

This strategy allows the use of the same intermediates as in the cold synthesis. Most often, however, the intermediate and the synthesis must be modified by building in a functional group that will accept the label.

For example, a double bond can be tritiated (Figs. 7.11 and 7.12), and a triple bond can be tritiated (Fig. 7.13) (11,12). An aromatic halide can be metallated then acidified with tritium (Fig. 7.14) (13). An aromatic halide can be catalytically reduced with tritium (Figs. 7.15–7.17) (14,15). An intermediate can be oxidized then reduced with tritide (Figs. 7.18 and 7.19) (16,17). A drug substance can be demethylated then methylated (Fig. 7.20) (18) or the labeled methyl can be introduced in the last step of the synthesis (Figs. 7.21–7.23) (9,19,20). Such methylation strategy has the advantage that several different labels can be introduced (Fig. 7.24) (5).

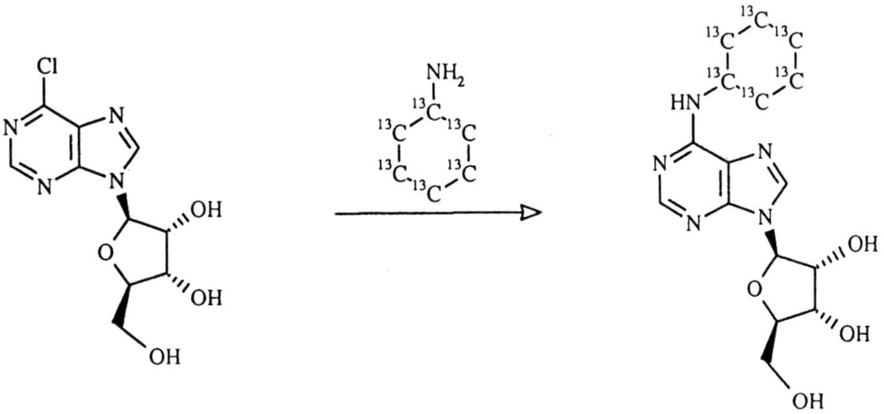

Fig. 7.6. Six ^{13}C labels.

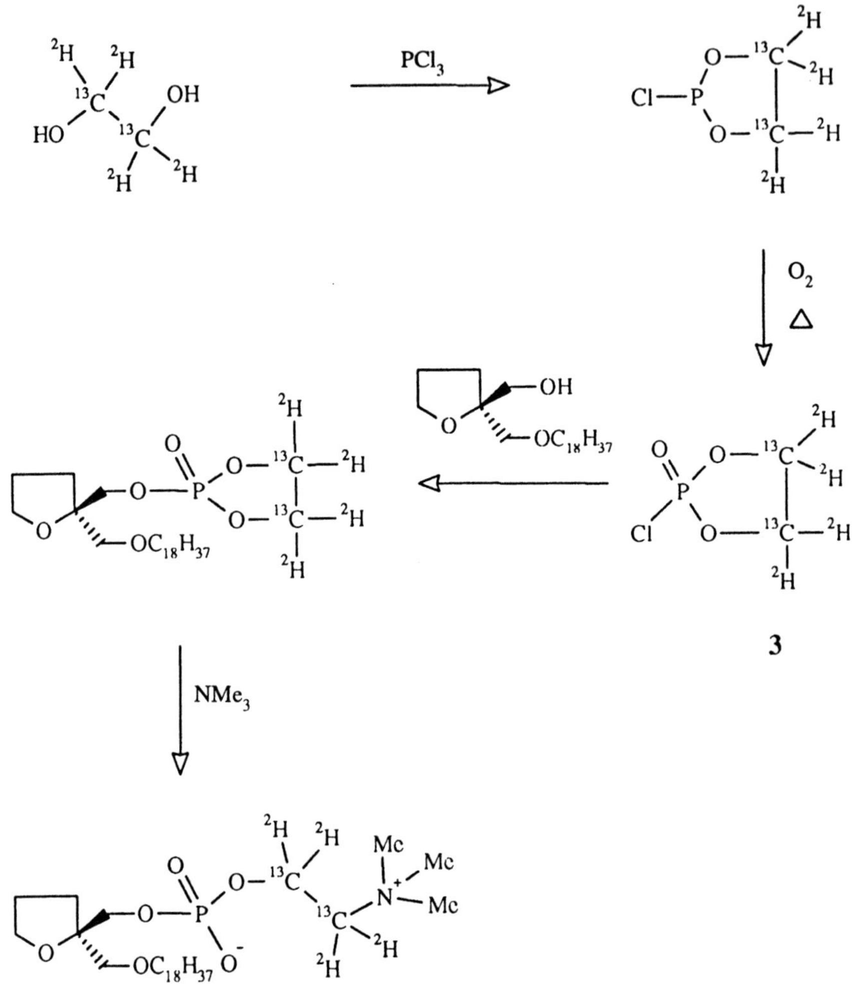

Fig. 7.7. Choline phosphate with two stable labels.

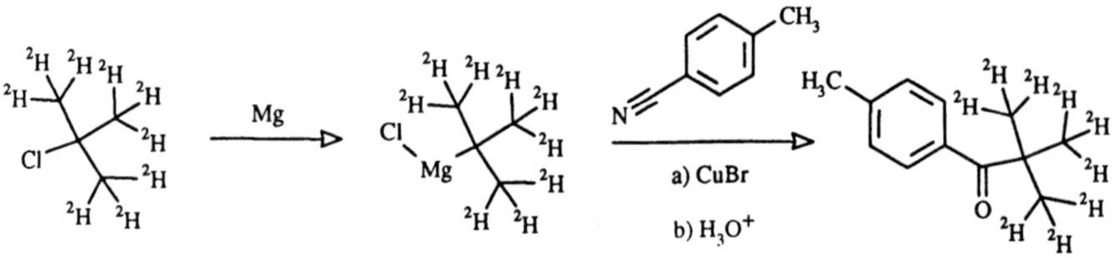

Fig. 7.8. Deuterated porcupine.

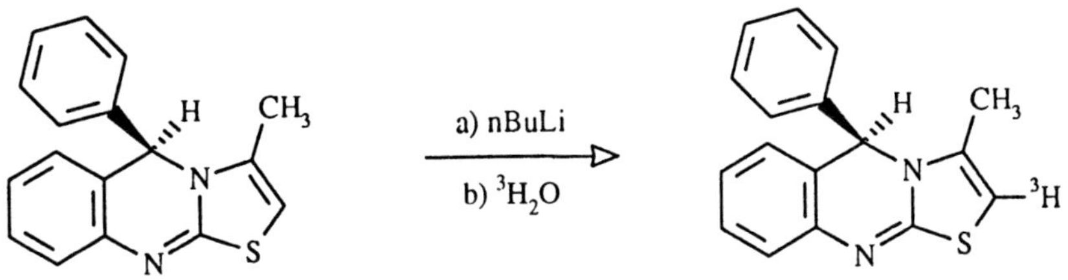

Fig. 7.9. Tritiation of the most acidic position in a thiazole.

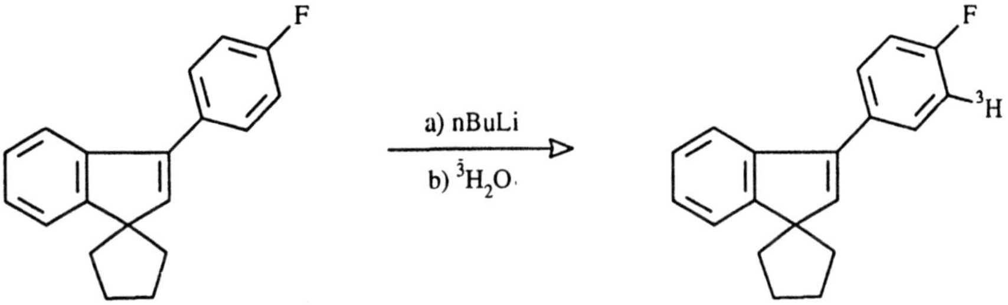

Fig. 7.10. Tritiation *ortho* to fluorine.

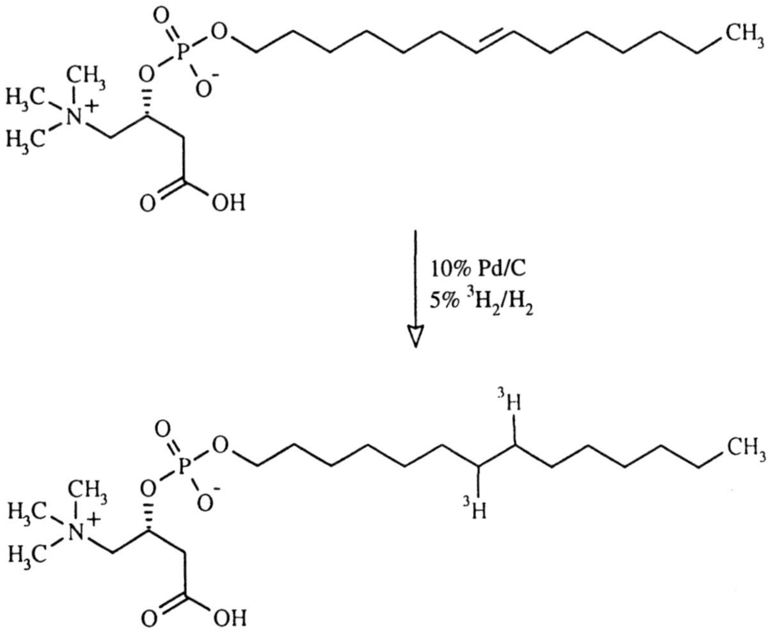

Fig. 7.11. Tritiation of a double bond.

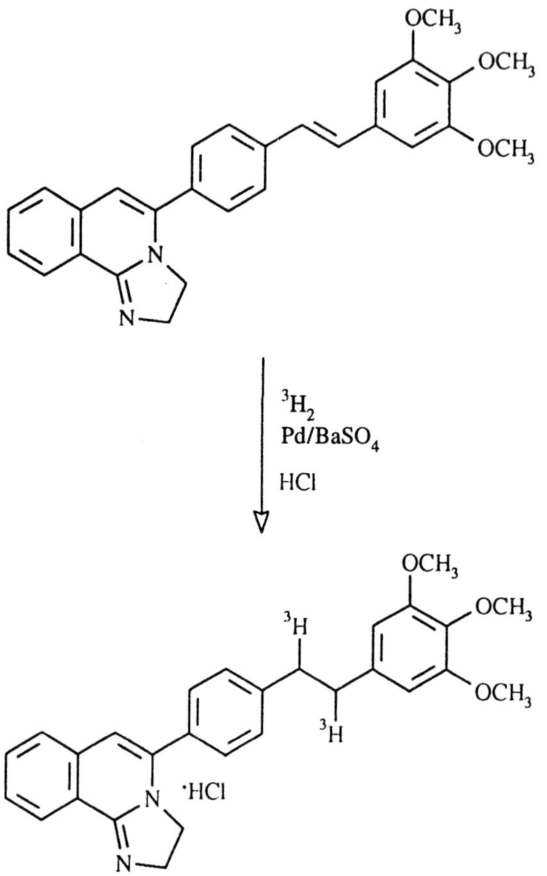

Fig. 7.12. Tritiation of a double bond.

Tritiation can also be nonselective. For example, by electrophilic substitution on benzene, [³H]sulfuric acid spreads tritium over all available positions, as in the example of labeling clemastine (Fig. 7.25) (21).

7.4. C-LABELS

As we have seen, ¹³C-labels are also conveniently incorporated by methylation or by esterification with (¹³C)isopropanol, obtained by reducing (¹³C)acetone with NaBH₄ (Fig. 7.26) (22).

Similarly, methylations with [¹⁴C]methyl iodide are an efficient way of introducing ¹⁴C (Fig. 7.27) (23). Amino acids can be labeled by methylation in an indirect way, by formylation and reduction, as in this example of Fmoc-[N-¹⁴CH₃]leucine, used to label cyclosporins (Fig. 7.28) (24).

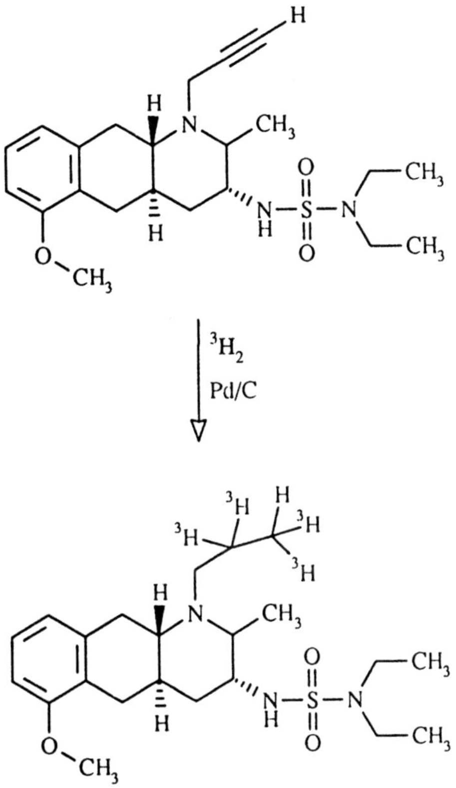

Fig. 7.13. Tritiation of a triple bond.

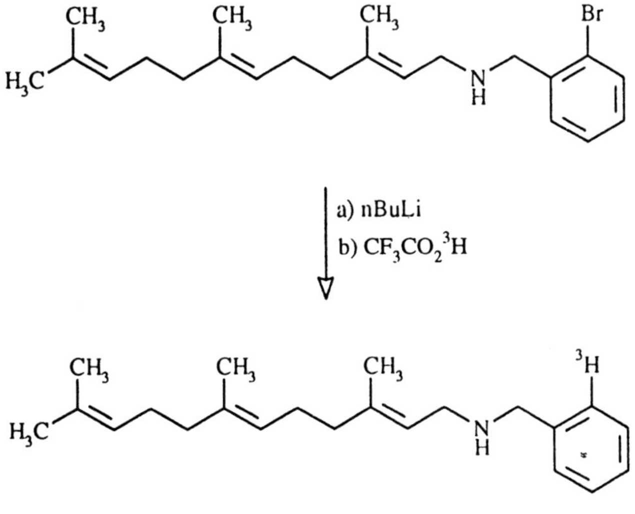

Fig. 7.14. Metallation and acidic tritiation of a halide.

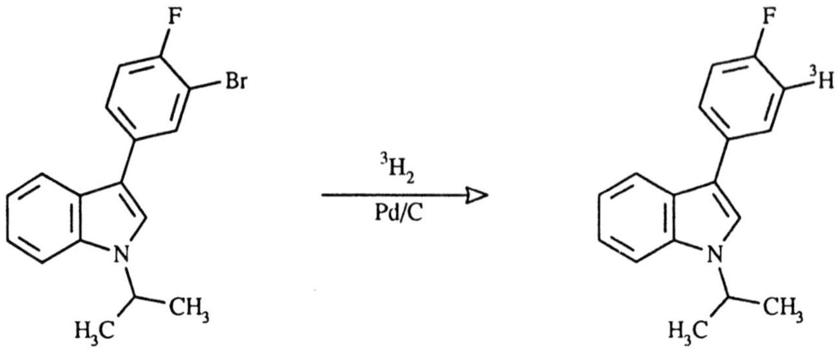

Fig. 7.15. Catalytic tritiation of an aromatic halide.

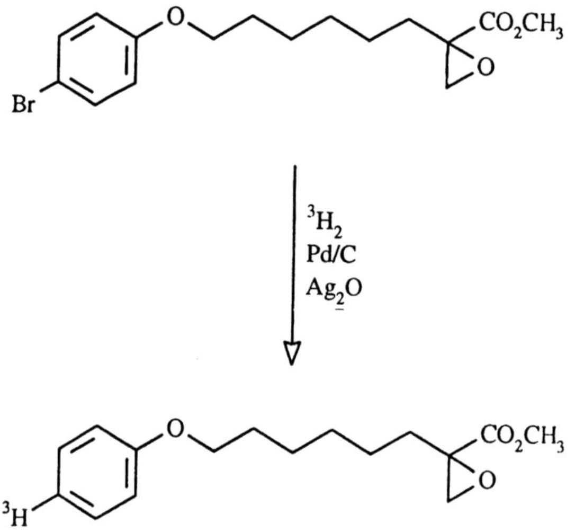

Fig. 7.16. Catalytic tritiation of an aromatic halide.

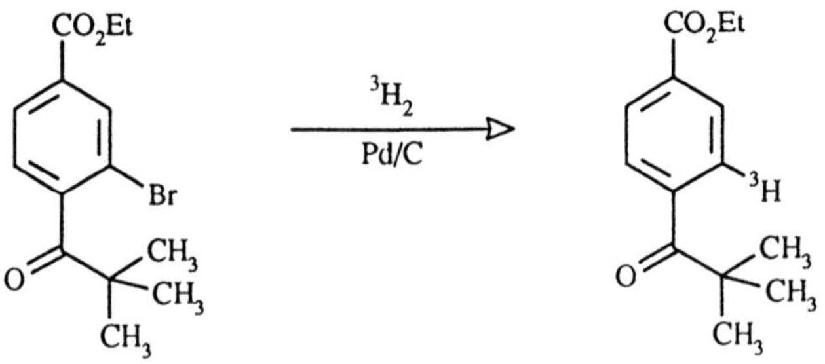

Fig. 7.17. Catalytic tritiation of an aromatic halide.

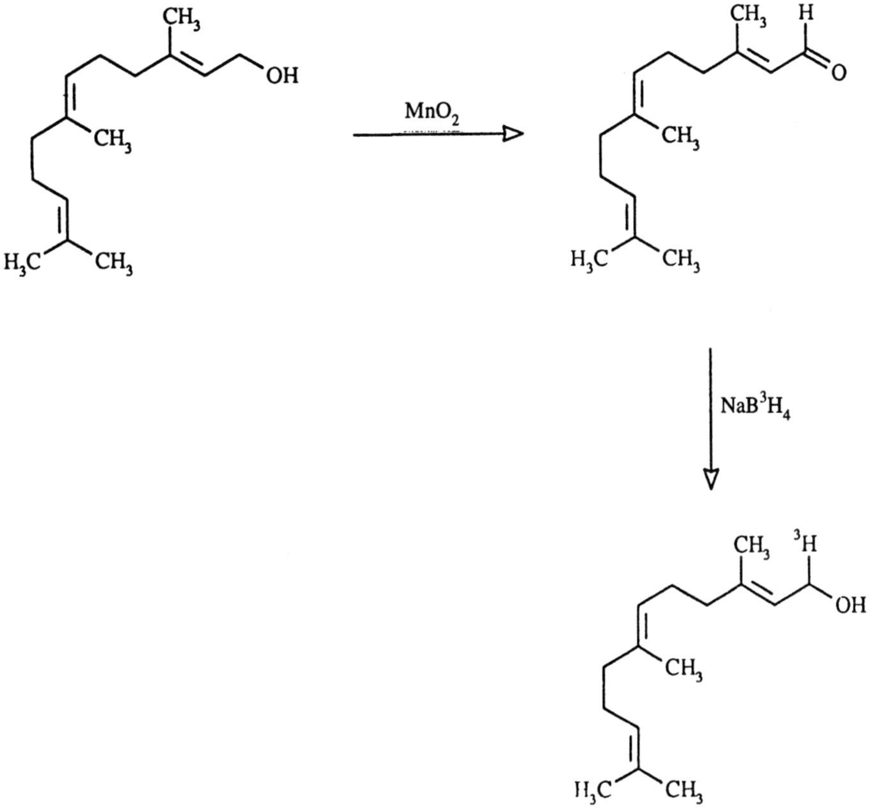

Fig. 7.18. Tritium labeling of farnesol.

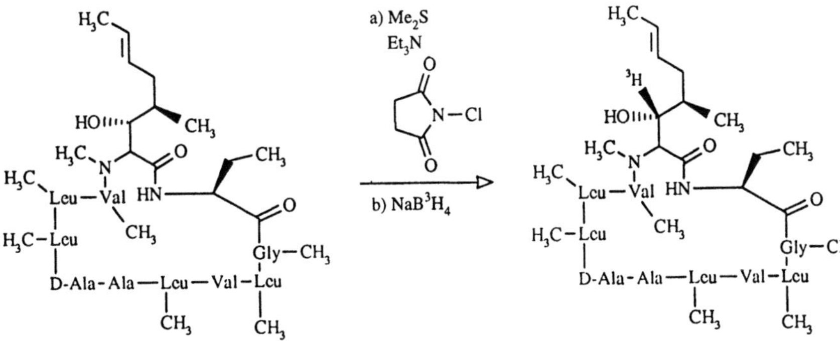

Fig. 7.19. Tritium labeling of cyclosporin.

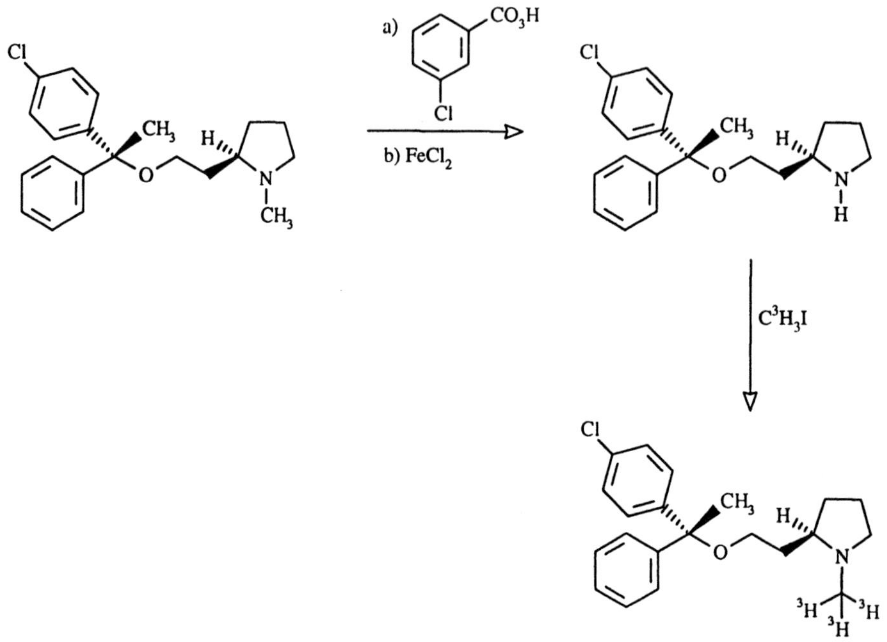

Fig. 7.20. Tritium labeling of clemastine.

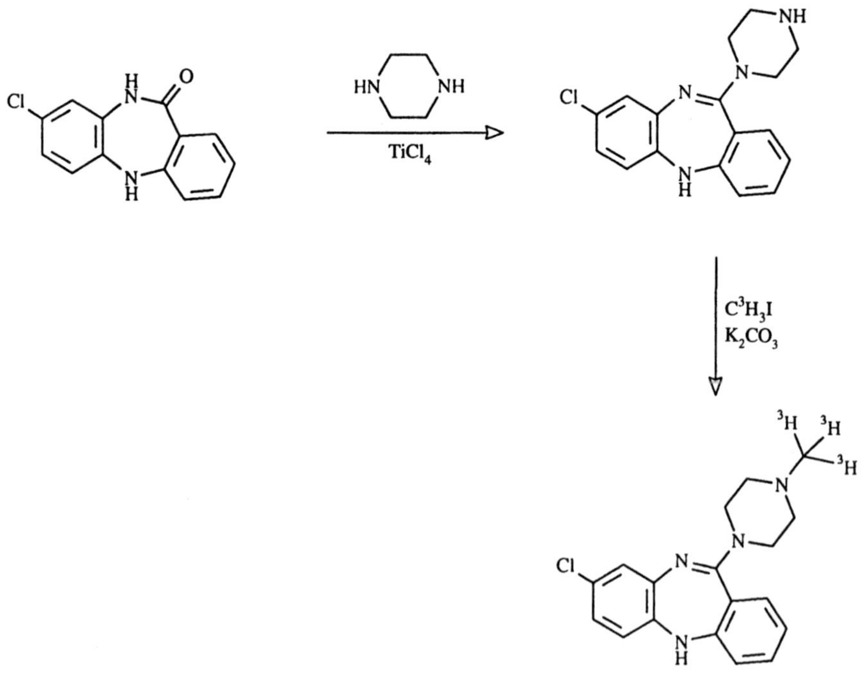

Fig. 7.21. Tritium labeling of clozapine.

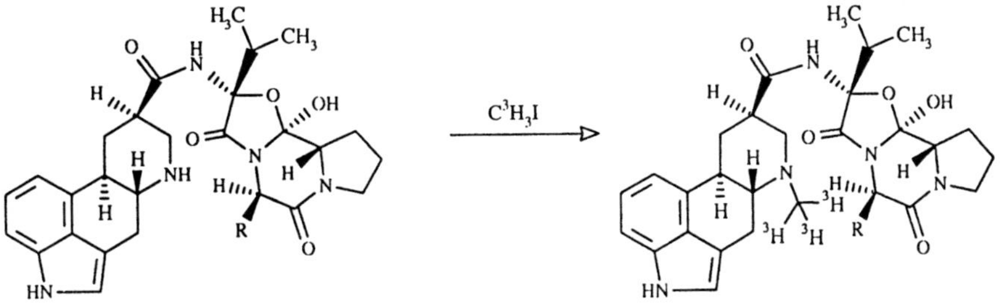

Fig. 7.22. Tritium labeling of hydergine.

C-labeled compounds usually require original and longer syntheses, however, as the carbon must be built into the backbone of the molecule using small labeled raw materials. Available sources of ^{14}C are, e.g., [^{14}C]barium carbonate, [^{14}C]carbon dioxide, [^{14}C]cyanide, [^{14}C]bromoacetyl chloride, and [^{14}C]malonate. The following examples will illustrate the use of these reagents.

A labeled carboxylate function can be added to a benzene ring either with $^{14}CO_2$ via metallation of a halide (Fig. 7.29) (25), or with $^{14}CN^-$ via a Rosenmund–von Braun reaction, which is followed either by a hydrolysis (Fig. 7.30) or by an addition of a Grignard reagent (Fig. 7.31) (7,9). In Figure 7.29, the carboxylate was condensed to form the central ring of the drug substance; in Figure 7.30, the carboxylate was reduced to become the methyl group in the product.

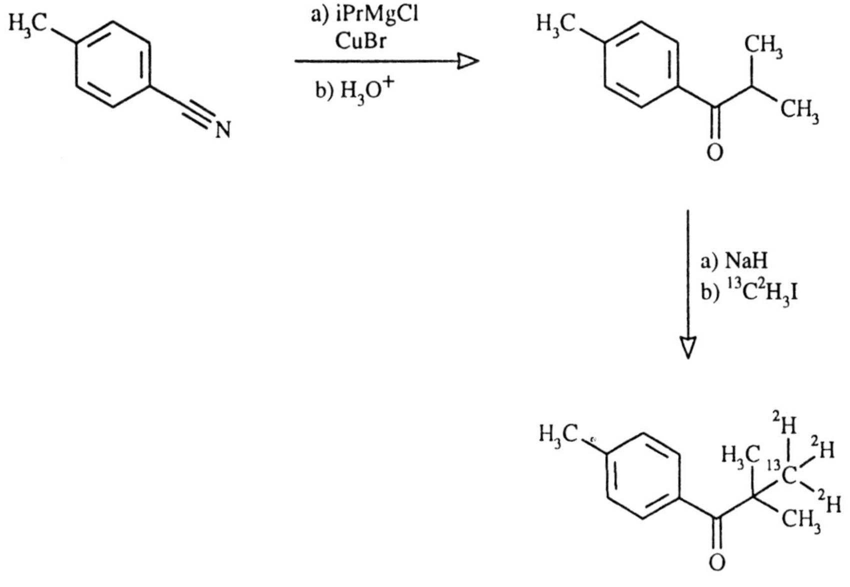

Fig. 7.23. Methylation with stable isotopes.

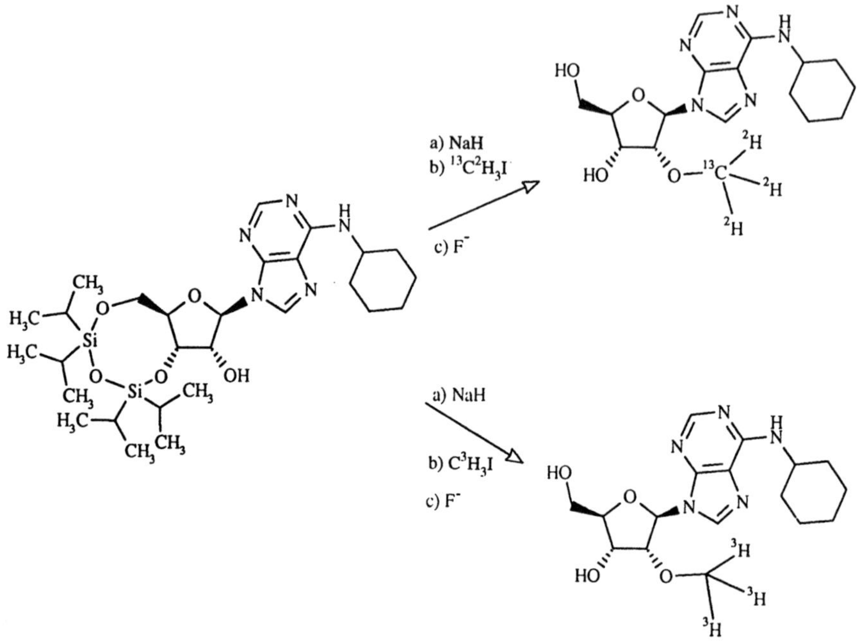

Fig. 7.24. Methyl iodide with different labels, stable or radioactive.

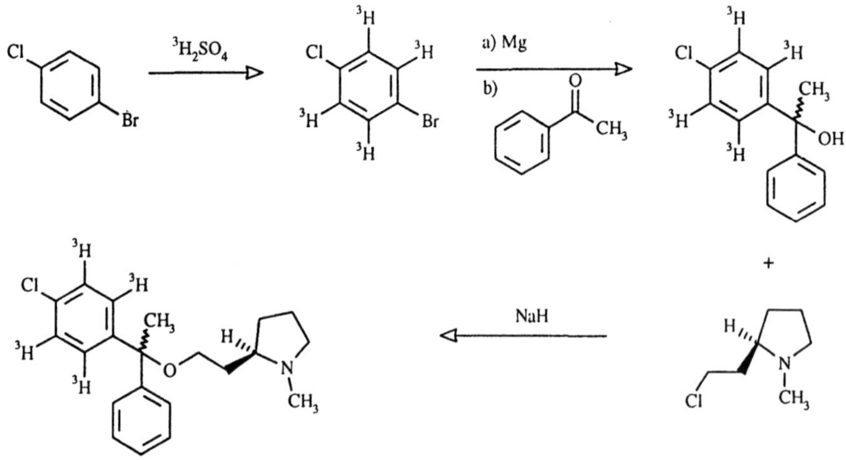

Fig. 7.25. Nonselective tritiation of benzene with $^3H_2SO_4$.

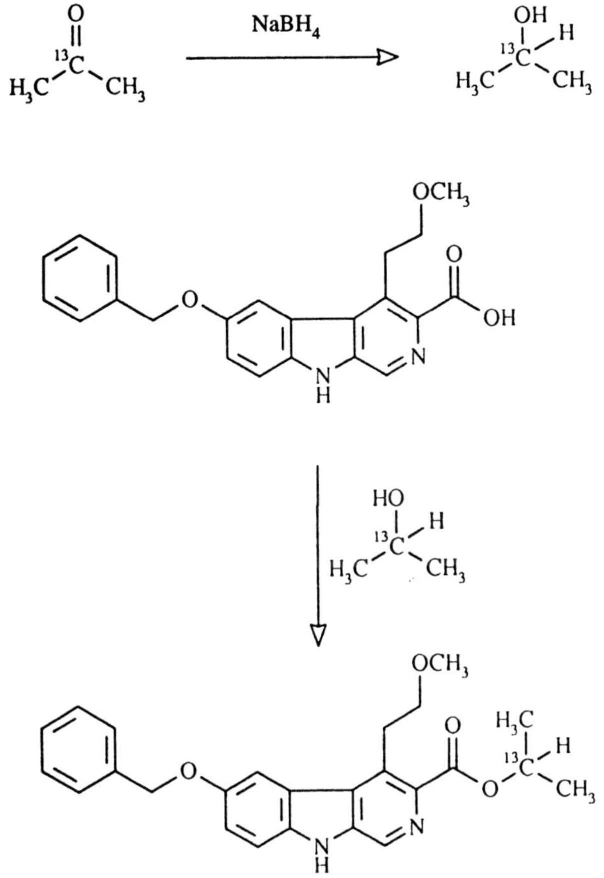

Fig. 7.26. Esterification with (^{13}C)isopropanol.

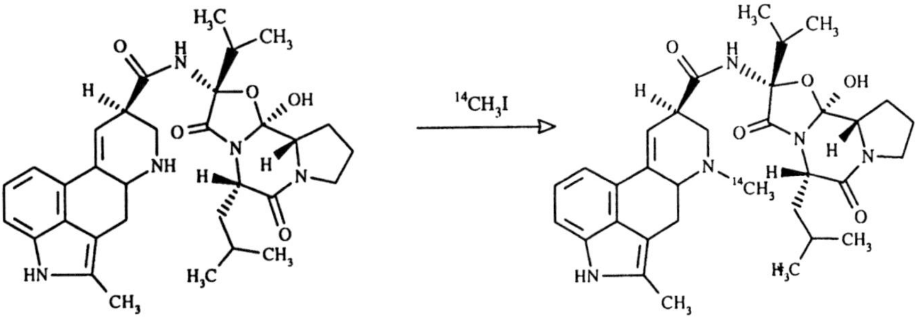

Fig. 7.27. Methylation with $^{14}CH_3I$.

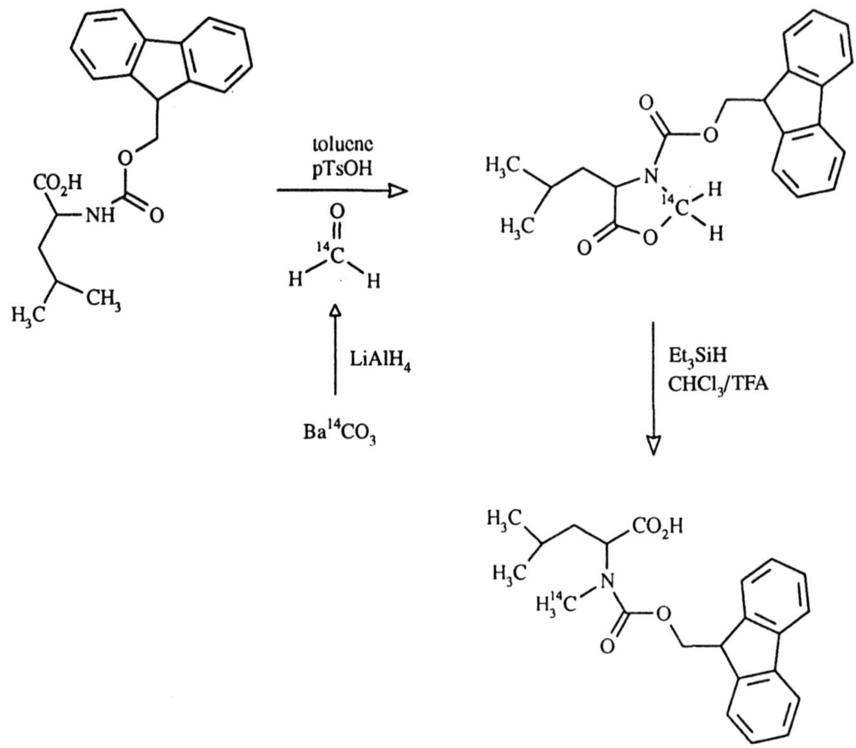

Fig. 7.28. Labeling of N-methyl amino acids.

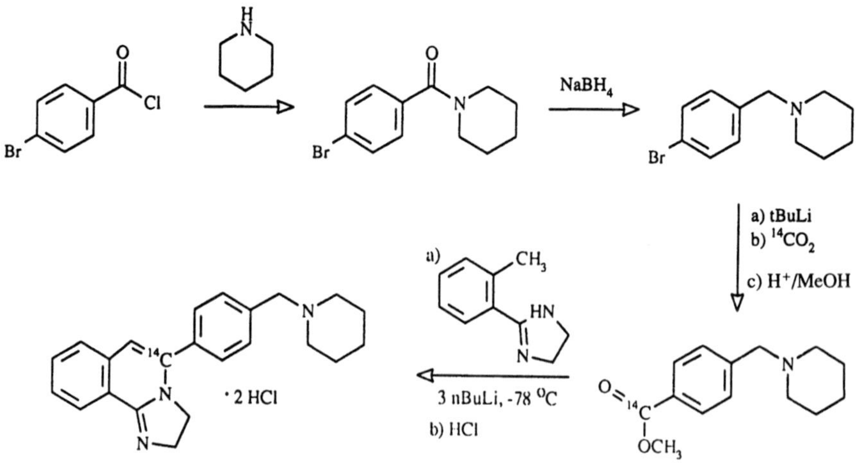

Fig. 7.29. Metallation and addition of $^{14}CO_2$.

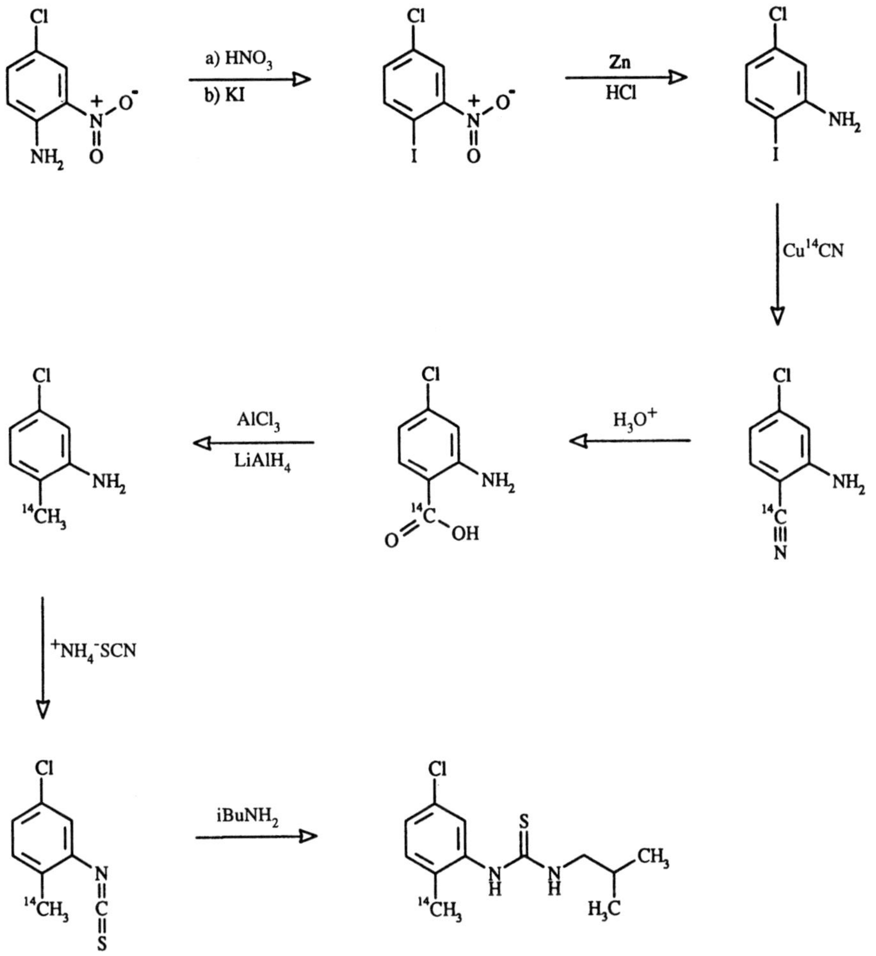

Fig. 7.30. Rosenmund–von Braun reaction followed by hydrolysis.

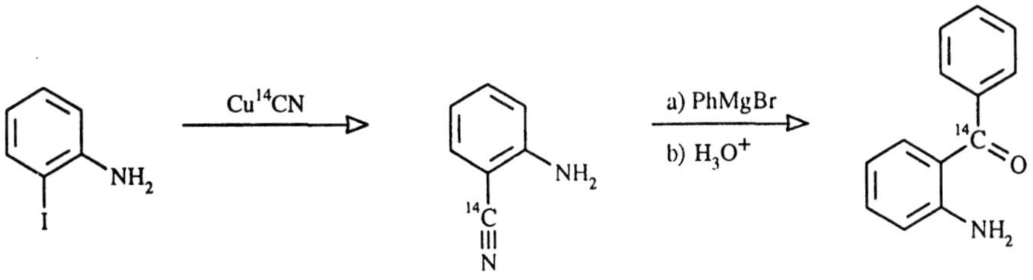

Fig. 7.31. Rosenmund–von Braun reaction followed by a Grignard addition.

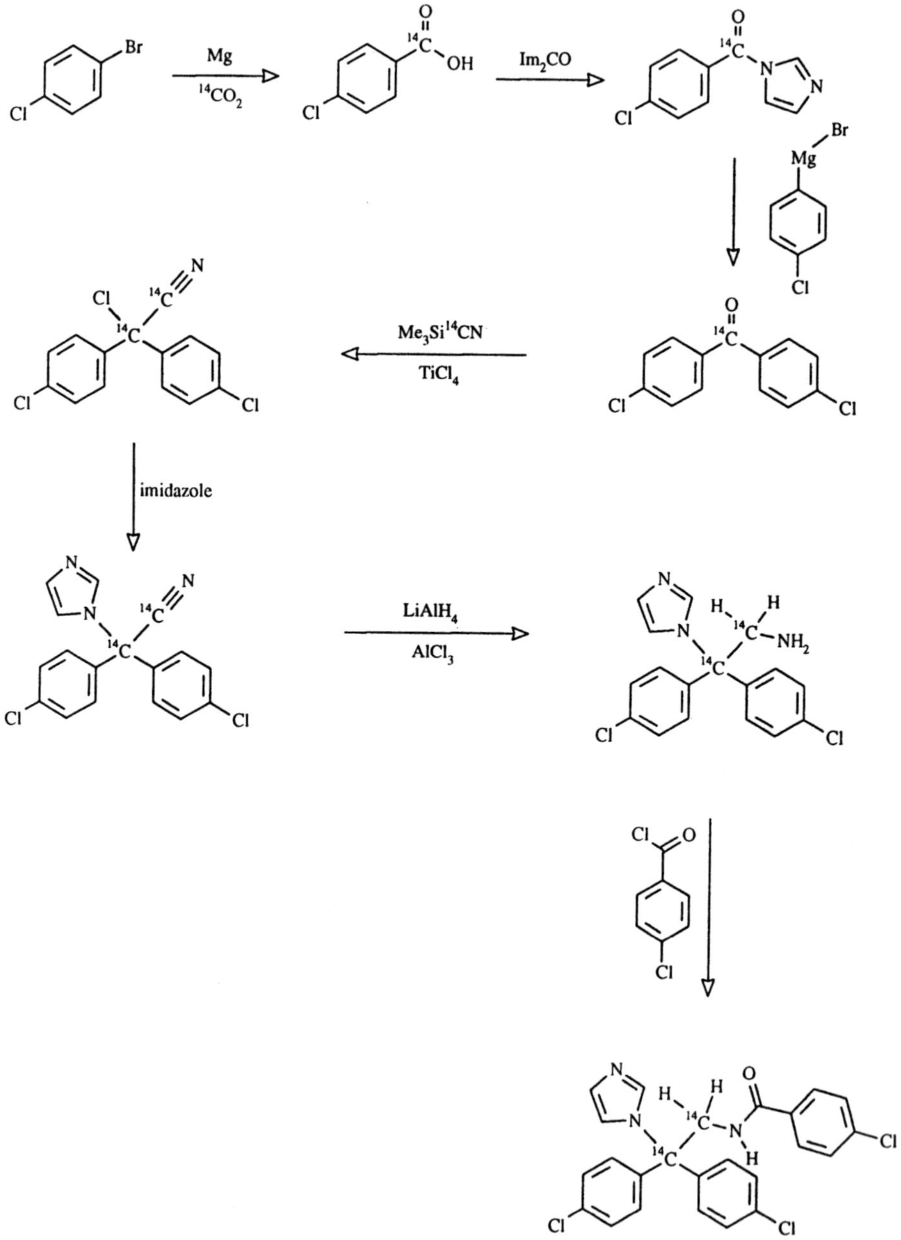

Fig. 7.32. Two ^{14}C labels in one molecule.

Sometimes both reagents ($^{14}CO_2$ and $^{14}CN^-$) can be used in the same synthesis (Fig. 7.32) (26). By choosing suitably labeled or unlabeled reagents (CO_2 and Me_3SiCN), one or the other of the two carbons was labeled, or both. Instead of placing both labels in one molecule, it is possible to mix two differently labeled compounds to achieve the same result. The choice of labeling position will finally depend on the stability of the product and the metabolic degradation pathway that might extrude the label. Similarly, clozapine was labeled with $Cu^{14}CN$ (Fig. 7.33) (19).

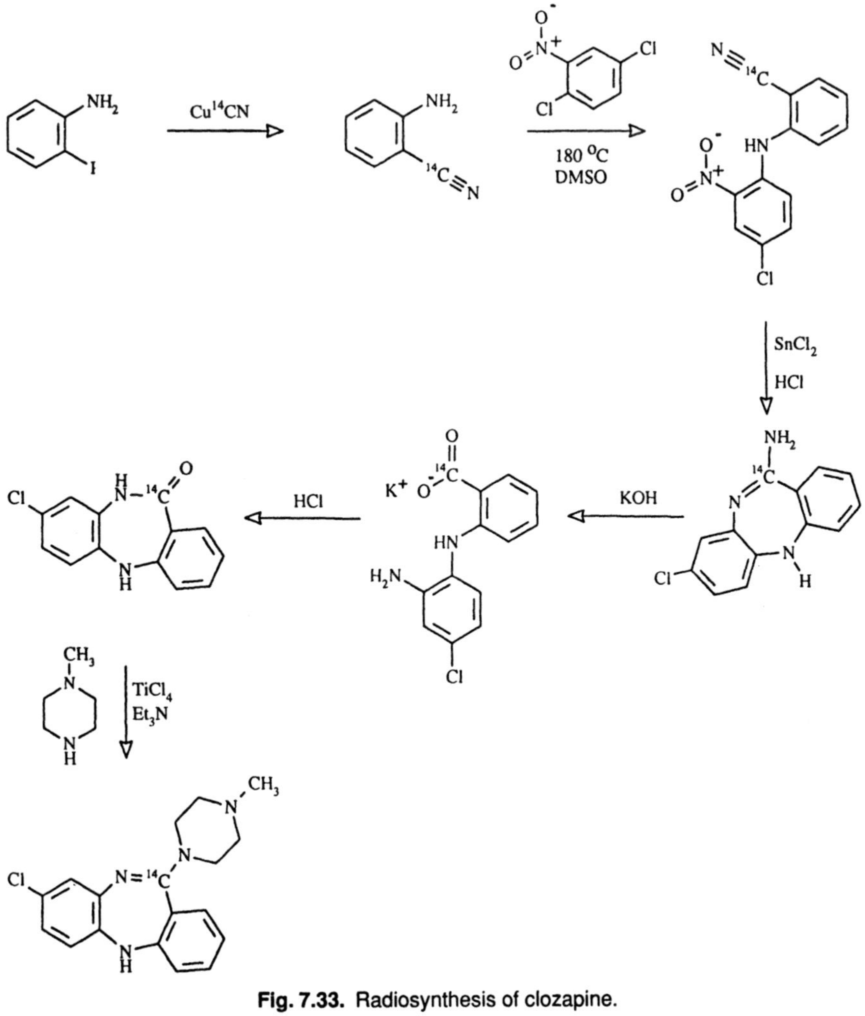

Fig. 7.33. Radiosynthesis of clozapine.

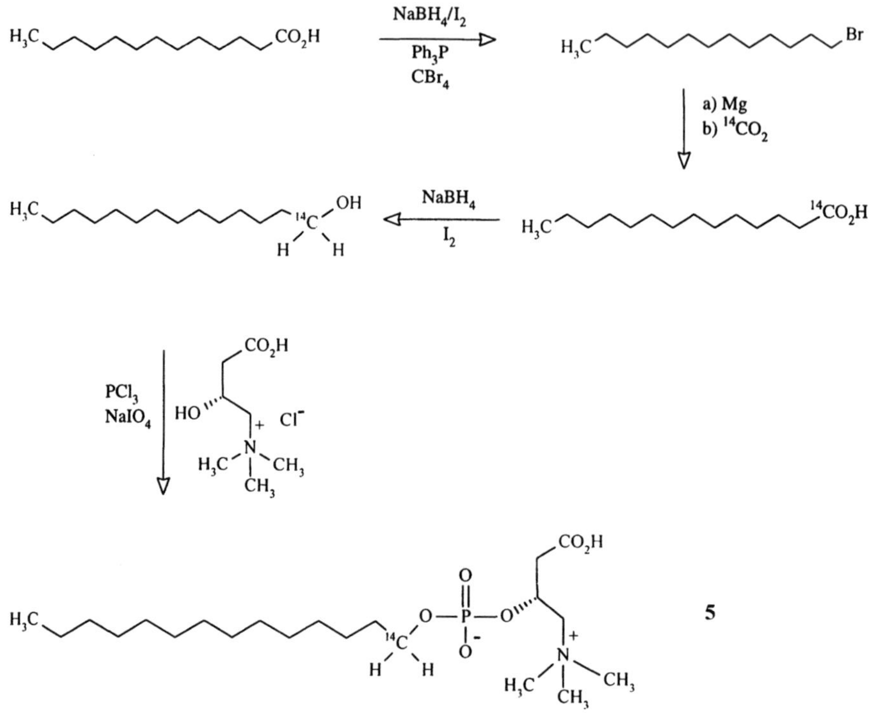

Fig. 7.34. Radiosynthesis of myristic acid and phospholipid **5**.

Myristic acid was similarly labeled via a Grignard reaction with $^{14}CO_2$ and used in the synthesis of phospholipid **5** (Fig. 7.34) (27). Stearic acid was labeled with [^{14}C]potassium cyanide and used in the synthesis of **6**, a platelet-activating factor analog (Fig. 7.35) (28).

Fluvastatin (discussed in Chapter 3) was labeled with [^{14}C]bromoacetyl chloride in the first step of the synthesis to bury the label deeply in the skeleton of the molecule, in position 3 of the indole (Fig. 7.36) (29). Similarly, labeling of the platelet-activating factor analog **7** inside the ring, where it will not be lost to metabolism, required an original total synthesis that started with [^{14}C]malonate (Fig. 7.37) (30).

This chemistry radiates excitement, not only β-particles!

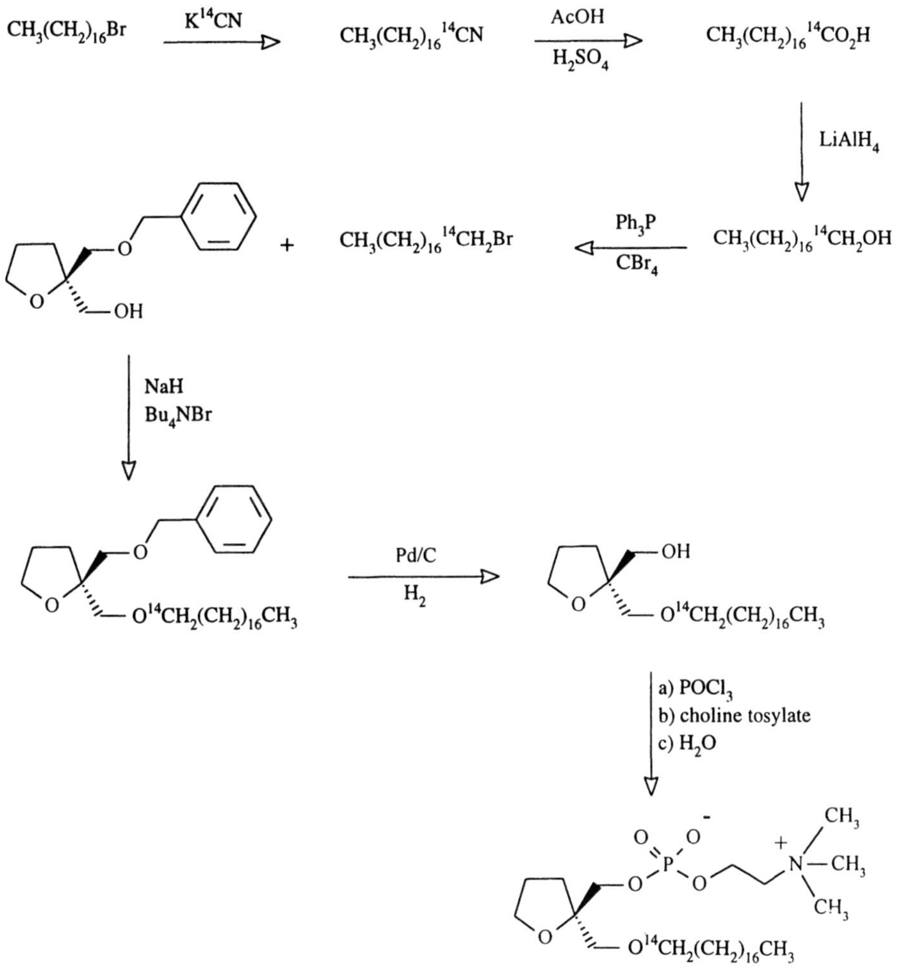

Fig. 7.35. Radiosynthesis of stearic acid and phospholipid **6**.

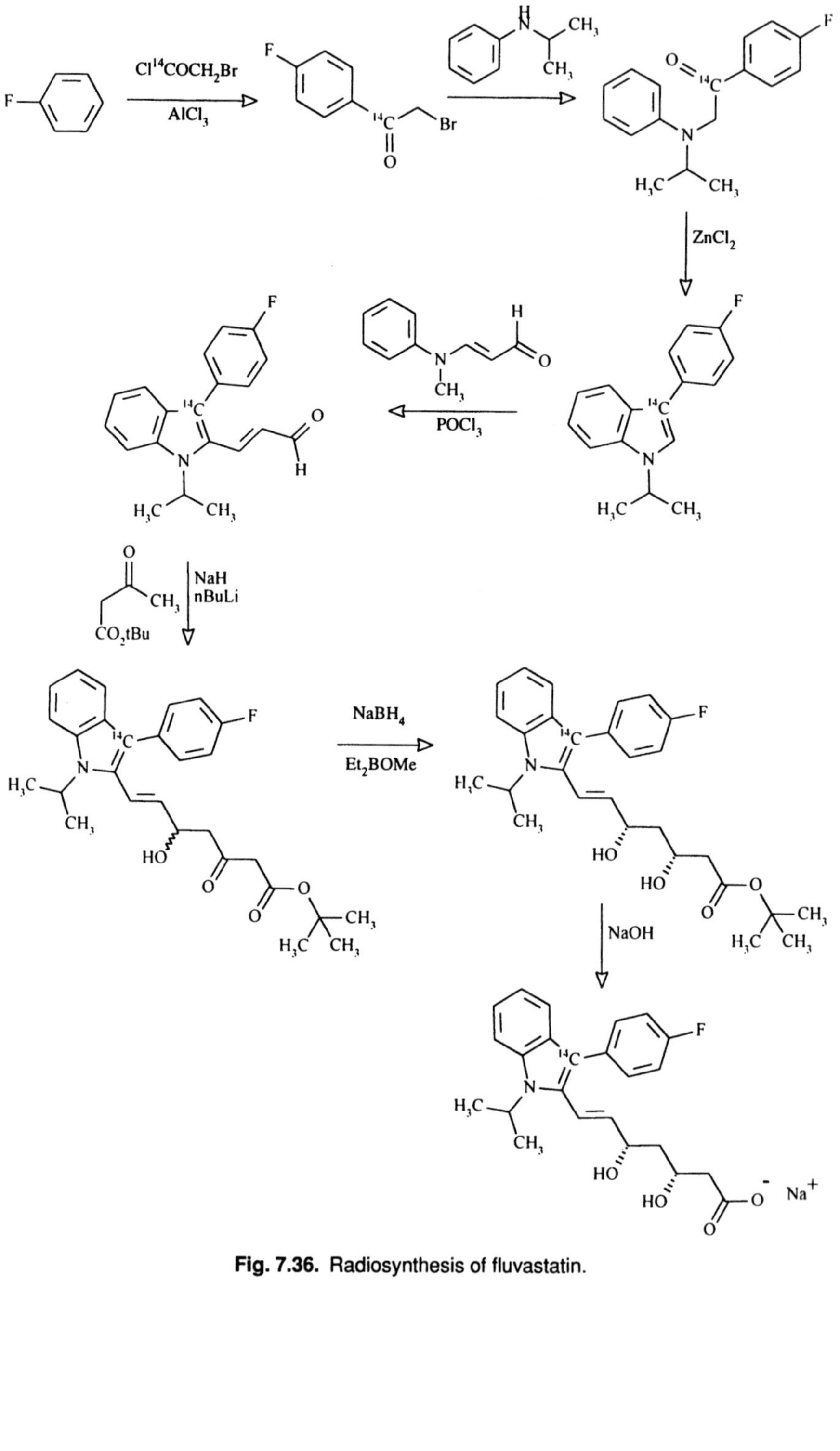

Fig. 7.36. Radiosynthesis of fluvastatin.

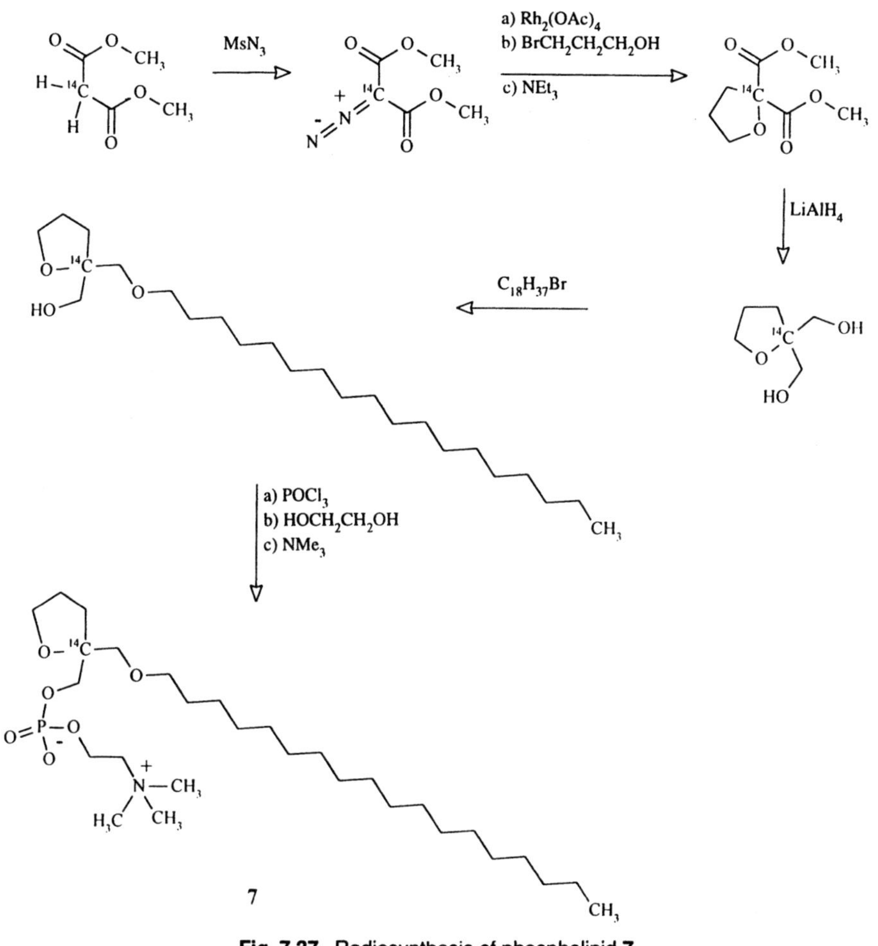

Fig. 7.37. Radiosynthesis of phospholipid **7**.

BIBLIOGRAPHY

1. L. Jones, R. Tarapata, and U. Sunay, personal communication, 1992.
2. M. J. Connor, J. A. Blair, and H. Said, "Secondary Isotope Effects in Studies Using Radiolabelled Folate Tracers," *Nature,* **287,** 253–254 (1980).
3. G. Ciszewska and U. Sunay, personal communication, 1993.
4. K. Talbot and U. Sunay, personal communication, 1993.
5. S. Ostrander and U. Sunay, personal communication, 1992.
6. S. Hathaway and U. Sunay, personal communication, 1991.
7. G. Ciszewska, K. Talbot, and U. Sunay, personal communication, 1992.
8. S. Hathaway and U. Sunay, personal communication, 1993.
9. V. Galullo, K. Talbot, and U. Sunay, personal communication, 1991.

10. K. Talbot, and L. Jones, personal communication, 1991.

11. A. Susàn, and K. Talbot, personal communication, 1989.

12. A. Susàn, K. Talbot, and V. Galullo, personal communication, 1987.

13. A. Susàn and U. Sunay, personal communication, 1991.

14. G. Conway and L. Jones, personal communication, 1988.

15. J. Bajwa and U. Sunay, personal communication, 1993.

16. F. Tang and A. Susàn, personal communication, 1988.

17. T. Mönius and co-workers, "H-3 Labeling of Cyclosporin in the *N*-methyl-γ-butenyl-γ-methylthreonine Moiety at High Specific Activity," in E. Buncel, G. W. Kabalka, eds., *Synthesis and Applications of Isotopically Labeled Compounds,* Elsevier, New York, 1992, 149–154.

18. V. Galullo, K. Talbot, L. Jones, and A. Susàn, personal communication, 1987.

19. U. B. Sunay, K. C. Talbot, and V. Galullo, "Synthesis of Carbon-14 and Tritium Labeled Analogues of the Novel Antischizophrenic Agent Clozapine," *J. Labeled Compounds Radiopharm.* **31,** 1041–1047 (1992).

20. H. Andres and co-workers, "A New and Efficient Synthesis of Monotritiomethyl Iodide," *J. Labeled Compounds Radiopharm.,* **27,** 767–776 (1989).

21. Y.-S. Tang and U. Sunay, personal communication, 1987.

22. T. Ray and U. Sunay, personal communication, 1993.

23. F. Tang and V. Galullo, personal communication, 1988.

24. R. M. Freidinger, J. S. Hinkle, D. S. Perlow, and B. H. Arison, "Synthesis of 9-Fluorenylmethyloxycarbonyl-Protected N-Alkyl Amino Acids by Reduction of Oxazolidinones," *J. Org. Chem.,* **48,** 77–81 (1983).

25. Y.-S. Tang and U. Sunay, personal communication, 1991.

26. B. Levine, and A. Susàn, personal communication, 1990.

27. G. Ciszewska and U. Sunay, personal communication, 1992.

28. G. Ciszewska, K. Talbot, and U. Sunay, personal communication, 1993.

29. Y.-S. Tang, L. Jones, and U. Sunay, personal communication, 1991.

30. F. R. Kinder, Y.-S. Tang, and U. B. Sunay, "Synthesis of Carbon-14 and Tritium Analogues of the Phospholipid Antitumor Agent SDZ 62-834 zi," *J. Labeled Compounds Radiopharm.,* **31,** 829–835 (1992).

FURTHER READING

R. Voges, "Stereoselective Procedures in the Synthesis of Enantiomerically Pure Isotopically Labeled Compounds," in J. Allen and R. Voges, eds., *Synthesis and Applications of Isotopically Labeled Compounds,* John Wiley & Sons, Inc., New York, 1995, pp. 1–26.

8

LICENSE*

The pharmaceutical industry is perhaps the most regulated industry in the United States. The U.S. Food and Drug Administration (FDA) has the power to stop the sale of any pharmaceutical if it considers it unsafe, unreliable, or defective. Sometimes lack of proper documentation may trigger the implication that production is not under control and, therefore, that the product is possibly defective. In this chapter I will address how the FDA and similar European agencies affect chemical development; as we will see, the consequence is more meetings, more documentation and more experimentation, especially in the area of process validation.

The FDA regulations were originally written for drug products (galenical process development, formulation scale up, and dosage production functions) but are now being applied frequently to drug substances (process development laboratories, pilot plants, production plants) (1). Present FDA guidelines attempt to explain how these regulations apply to drug substances (2). For example, the term *process validation* was developed with drug products in mind; the industry and the FDA have to agree on what it means when applied to drug substance, because the term (but, as we will see, not the activity) was foreign to chemical development until recently. This chapter describes one interpretation and one point of view—from the industrial perspective—of the FDA guidelines, summarizing the good manufacturing practices (GMPs), describing what is contained in a development report, and defining drug substance process validation.

*Based on lectures I gave at the International Exhibition and Conference on Pharmaceutical Ingredients and Intermediates, Torino, Italy, Sept. 24, 1993, and at the Drug Information Association Meeting, Philadelphia, Pa., Feb. 7, 1992, published as "Effects of FDA Inspections on Chemical Development," *Drug Inform. J.*, **27**, 469–480 (1993).

The FDA guidelines intertwine different concepts (GMP, process development, process change control system, analytical control), but it is easier to understand these different activities if they are treated separately (Fig. 8.1).

For example, GMP is a production activity, characterized by constancy, reproducibility, high quality, and specifications passed. GMP relies on the analytical control system (which checks quality) at the end of the process, and on process development (which builds quality in) up front of the production. Process development, on the other hand, is characterized by making changes in the process (some of which will fail specifications); this activity will finally define the ultimate process and the ranges of critical process parameters and will thus guide GMP. For example, if production stays within the specified ranges, the product is expected to pass quality specifications. If these ranges are exceeded, the change control system must be employed, with its proscribed record-keeping, approvals, and additional analytical controls. Process development is also served by analytical control when defining process variables. To summarize, GMP, process development, analytical control, and the change control system are four totally different activities, but they all serve each other and depend on each other.

8.1. GOOD MANUFACTURING PRACTICES

GMP is a set of standards for producing drug products (2). The same standards also apply to the production of drug substances, including not only commercial production but also batches for clinical (human) tests and those used to support regulatory filings. Because clinical batches are often produced in chemical development, where development and production meet, this topic is brought up here. There is no general agreement on how many steps in a synthesis must follow GMP: all steps,

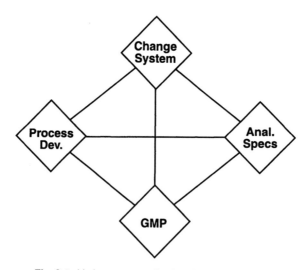

Fig. 8.1. Various systems in chemical development.

steps after the key intermediate (generally defined as any intermediate that can affect the quality of the drug substance), or only the steps after the last purification. This issue is cloudy because drug substance preparations differ from drug product preparations (for which the GMP rules were written) in one critical respect: Drug substance preparations include purification steps, and the quality of the earlier synthetic steps may not have any effect on the quality of the drug substance. Often the expenditure of the extra time and capacity needed for GMP is justified by saying, "GMP is just good science." Mostly this is true. However, one can also make it bad science. For example, if one applies GMP rules too rigidly, one is intimidated and will not make changes (in the process, in specifications, or in analyses), even when good science indicates changes are desirable.

The assumption behind these standards is that the existence and application of GMP ensure high-quality drugs. Here is a summary of GMPs.

8.1.1. Organization and Personnel

A separate and independent quality control unit must be established in the organization to test, approve, and reject products; set and approve specifications for raw materials and products; and write and approve these and other written procedures. The qualifications of the personnel in the manufacturing units must be documented with job descriptions and with curricula vitae. All personnel associated with production must be trained in GMP, and these classes must be documented with signatures of attenders.

8.1.2. Facilities

The manufacturing facilities must have adequate space, heating, ventilation, and plumbing and be constantly maintained to ensure high-quality production. Measures must be taken and documented to prevent cross-contamination between drugs or with extraneous materials. For parenteral drug preparation, aseptic rooms are required (a rating of $<100,000$ particles/ft^3 and <25 microbial units per 10 ft^3); this applies also to the last step and isolation of parenteral drug substances.

8.1.3. Raw Materials

All raw materials, solvents, and intermediates must be of suitable quality. Generally, this means that they must be sampled, analyzed, and approved based on preset and preapproved control procedures and specifications. Certificates of analysis are required (internal or from the supplier). Rejected materials must be physically separated and quarantined (again, to prevent mixups and contamination).

8.1.4. Equipment

Written operational instructions must be in place for all special equipment (e.g., a freeze-dryer). Standard operating procedures (SOPs) must be written for calibrating and maintaining all equipment. Procedures must be written for cleaning processing equipment. Avoid equipment that can contaminate the product (e.g., fiber-releasing filters).

8.1.5. Production Records

The manufacturing procedure must be well documented, signed, and dated. Any deviations from the process must be recorded and justified. Deviations are unintentional or one-time changes made in the process, as opposed to permanent process changes, which are handled by the process change control system (see below). Batch numbers must be assigned and recorded for each batch of material made. Weights and yields must be recorded in the batch record and countersigned. All components must be released by the quality control unit. Theoretical and actual yields must be calculated and recorded. In-process controls and specifications must be written down. Any reprocessing procedures must be preapproved in writing.

8.1.6. Distribution and Storage

Written procedures must be in place for warehousing, quarantine, and distribution of lots of drug substance and product. Storage conditions must be prescribed (based on stability tests) and recorded.

8.1.7. Packaging and Labeling

Written procedures must exist for the receipt, identification, and storage of labels and packaging materials. Label issuance must be controlled. Labeling of products must be inspected and the inspection recorded. Drug substances and products require expiration dates (based on stability tests).

8.1.8. Analysis

Analytical methods and specifications must be established for raw materials, intermediates, and drug substances. All deviations and changes must be approved and recorded. All instruments must be calibrated. Sampling procedures must be described in writing. Statistical methods for quality control should be used. Test methods must be validated (for sensitivity, specificity, reproducibility). Stability testing is needed to determine storage conditions and expiration dates. Reserve samples of all drug substances must be preserved.

8.1.9. Records

Good records are needed for all of the above. To summarize: One must maintain distribution records; calibration and maintenance records; and production review records, including deviation reports, batch records, master production records, inventory records, analytical reports and records, equipment cleaning log, and files reviewing recalls and complaints.

8.2. PROCESS CHANGE CONTROL SYSTEM

Unlike commercial production, in chemical development, process changes are the rule and not the exception. That is why the FDA requirement for process change control system is especially difficult for chemical development, both practically and philosophically. Nevertheless, all significant process changes are recorded, justified, and approved in one or several documents: in project review meeting minutes, in laboratory development reports, in laboratory process descriptions, in plant master procedures, in batch records, in batch experience reports, in plant development reports, and so on. In other words, we do not have one document describing process changes; many documents deal with process changes, which is the main function of chemical development. What makes a process change significant? Any change that can, in principle, change the quality of the product must be considered a significant process change (3,4). A few examples are a change in reaction and workup ingredients, of in-process control procedures, in specifications, in scale, in purification method, in the reaction temperature (beyond the prescribed range), in the order of addition of reactants (several examples in Chapter 2 point this out), or in the ratio of reactants and solvents. A new raw-material supplier (a common source of new impurities) is also a significant change.

8.3. THE DEVELOPMENT REPORT

The FDA inspection guidelines are reasonable in that they ask for good science and good development practices, i.e., a good understanding of the process; although most of the data are already available in good process development laboratories, these data are perhaps spread out over notebooks, in batch records, or even in chemists' heads. Today, a lot more documentation is required, because the data must be organized and packaged in easily presentable and easily understood ways, e.g., in development reports.

The FDA would like to see written evidence that the process is well understood and under control, because it believes that a well-developed process ensures a reproducible quality of the drug substance. The development report is the backbone for this written evidence and is just a summary of the history of process development, both in the laboratories and in the plants. The Sandoz development report, for example, contains these sections:

- Process scheme.
- Comparison of development processes.
- Comparison of development and production batches.
- Comparison of production batches.
- Explanation of each operation in the NDA process.
- Development history.
- Process validation.

Table 8.1. Comparison of Four Processes in Terms of Process Variables

Variable	Process			
	A	B	C	D
Reaction temperature, °C				
20 → 25	•			
25 → 30			•	•
30 → 35		•		
Reaction solvent				
MeOH	•	•		
THF			•	
EtOH				•
Extraction solvent				
tBuOMe	•		•	•
EtOAc		•		

The process scheme (see Fig. 1.2, Chapter 1) is generally a one-page diagram of the chemistry involved in the final synthetic process.

The next three sections of the development report deal with the reproducibility of the process. We document process changes that were made and show that all the processes are reproducible, in spite of any process changes. The comparison of development processes covers different development processes in terms of the process variables (Table 8.1).

For example, process A used a reaction temperature of 20 → 25°C, the reaction solvent was methanol, and the extraction solvent was t-butyl methyl ether. Process B was performed at 30 → 35°C, in methanol, and extracted with ethyl acetate, and so on for all other processes that had been used for production of critical batches. All development batches of drug substance are also compared in terms of the results (analyses), to show that they are under control and that they reproducibly pass specifications.

This comparison is again best done in tabular form (Table 8.2) in which the analytical controls, the specifications, and the results for each batch of drug substance are listed. Such a spreadsheet can, of course, have 50 lines depending on how many items are analyzed, and 100 columns, depending on how many batches were produced. Table 8.2 is an excerpt of a much larger table, just large enough to illustrate the point. Note three aspects: all of the runs are reproducible, all these products pass specifications, and the specifications are relatively wide since we are in

Table 8.2. Comparison of Development Batches

Specifications	Lot 32	Lot 33	Lot 34
Product, 97–102%	99.3	99.6	98.5
By-product 1, <3%	1.3	1.4	1.2
Total unknowns, <1%	0.3	0.7	0.7
pH, 8.0–9.5	8.4	8.5	9.2
Water, <4%	1.8	1.3	2.7

Table 8.3. Comparison of Development and Production Batches

Specifications	Lot 84	Lot 91911
Product, 98–102%	99.4	99.9
By-product 1, <1%	0.3	0.3
Total unknowns, 0.5%	Not detected	Not detected
pH, 8.0–9.5	8.7	8.8
Water, <4%	3.4	2.9

the early phase of development: e.g., by-product 1 may be present up to 3%, total unknowns up to 1%, and so forth.

A comparison of development and production batches is illustrated in Table 8.3, where the reproducibility on scale up is shown by comparing a late development batch with the first production-scale batch.

Again, note that the two are almost identical in quality and that by now, the specifications have been tightened, for example, to <1% of known and <0.5% of total unknown by-products, but actually the unknowns were no longer detected in these batches.

In a comparison of production batches (Table 8.4), at least three production-scale batches are compared to each other, to show reproducibility. The comparison of three production batches is one definition of the term process validation. There are other definitions, as we will see later. Again, notice how close the values are; this process is well under control.

The explanation of each operation in the new drug application (NDA) process helps the FDA reviewer or inspector. Here is an excerpt from an NDA:

Add to the solids 485 L (480 → 505 L) of ethyl acetate and heat the mixture with stirring to about 70°C (65 → 75°C) for 30 min (25 → 45 min) until complete solution results. Cool the mixture to 0°C (−2 → 2°C) and continue stirring at this temperature for 1 h (½ → 6 h).

This crystallization procedure, for example, is explained by saying that it "is performed to remove by-product 1." Note that we record ranges for parameters in the NDA procedure; these are the so-called verified ranges or experience ranges, i.e., the actual ranges that the plant has observed or used at one time or another. The significance of this will be seen later when we discuss process validation.

Table 8.4. Comparison of Three Production Batches

Specifications	Lot 91911	Lot 91912	Lot 91913
Product, 98–102%	99.9	100.3	100.3
By-product 1, <1%	0.3	0.4	0.4
Total unknowns, <0.5%	Not detected	Not detected	Not detected
pH, 8.0–9.5	8.8	8.9	8.8
Water, <4%	2.9	2.2	2.8

Table 8.5. Summary of all Reaction Conditions Tried

Reaction Conditions	Yield, %
$Pd(P\phi_3)_4$, CH_3CO_2H, CH_2Cl_2, 24°C	50
$Pd(OAc)_2$, $P\phi_3$, HCO_2NH_4, dioxane, reflux	10
$Pd(P\phi_3)_4$, HCO_2NH_4, dioxane, reflux	85
polymer-$Pd(P\phi_3)_4$, HCO_2NH_4, THF, reflux	85

The development history is another important and large section in the development report, which is also referred to as process justification. This section would describe the history of how the synthetic process was defined; e.g., it will show all syntheses tried, all reaction conditions tried, optimizations, and process changes. It is a summary or just a compilation of all research and development reports that we generate as we go along in process development. All syntheses that were developed or used during chemical development are again best summarized as synthesis schemes.

All reaction conditions that were tried during process development are listed with an optimization goal (Table 8.5), here the yield. Notice that in only a few experiments (by changing the reaction conditions) the yield could be raised to 85%. Another example of an optimization is in Table 8.6. Typically in optimization, one critical parameter is varied (in this example, the solvent ratio) to optimize an outcome (in this case, the isomer ratio). Note how the variable was fine-tuned to give a high stereoselectivity of 98%.

These are process variations tested before the process is defined; however, sometimes process changes, for a variety of reasons, also occur during or after batches of drug substances have been produced. Such changes during production deserve special attention. Significant process changes are not only recorded in reports to the FDA but also approved internally during development by the process change control system (see above).

The final section of the development report is process validation. The remainder of this chapter will concentrate on this topic and issue.

Table 8.6. Optimization of a Critical Variable—Solvent Ratio—to Maximize *syn*-Isomer

Solvent Ratio		Isomer Ratio	
THF	MeOH	*syn*	*anti*
20	1	94	6
10	1	96	4
6	1	97.5	2.5
4	1	98	2
3	1	97.6	2.4

8.4. DRUG SUBSTANCE PROCESS VALIDATION

8.4.1. Definition

The concept and topic of drug substance process validation are controversial, I believe, because the term is new to chemical development chemists (but it is well known to dosage formulation function, for which it was initially intended) (5). A clear definition, it seems, would quickly convince process chemists that it is not a strange activity for them, just a new word.

The FDA's guidelines on process validation define process validation as "establishing documented evidence that provides a high degree of assurance that a specific process will consistently produce a product meeting its predetermined specification and quality attributes." What does this statement mean? *Documenting* that a process *consistently* meets *specifications* is exactly what process validation is not, because using quality controls to prove reproducibility is exactly what the FDA does not consider sufficient (according to FDA speakers at meetings with the industry). Instead, one must build in quality and assurance of reproducibility up front, i.e., with process development.

Unfortunately, the FDA's inspection guide does not help in this respect either, as it offers multiple definitions for *process validation* (2). In various places in the guidelines, there are different definitions of process validation:

1. "Document that significant manufacturing processes perform consistently"; this definition is taken from the regulations and means that one must prove that the process is reproducible.
2. "Scale up and development reports, along with purity profiles, constitute the validation report"; this definition seems to imply the general activity we call process development.
3. "Limitations of a reaction . . . are usually identified in the development phase"; this definition can be interpreted to mean that one must find the limits of process variables.

Although these are all worthwhile and practiced, they constitute a spectrum of activities that are better called *process development,* a term used and understood by development chemists. The term *process validation* should refer only to the narrower activity of defining limits of process variables. We will come back to this proposed definition in a moment.

8.4.2. Drug Substance Versus Drug Product

There are two other aspects of drug substance process validation that make it controversial, I think. First, process validation for a drug substance is less critical than for a drug product, because analyses are more precise for substances. That is, drug substances are homogeneous, well-defined organic molecules and can be analyzed precisely and reliably (e.g., by high-pressure liquid chromatography, which quantifies the amounts of the by-products), whereas drug products are mixtures of different components (including silica gel, cellulose, and buffers) that are more difficult to analyze precisely. Second, purifications, which are an integral part of chemical syntheses, remove variability by removing by-products or impurities entering the synthesis and thus make the quality of the starting material less critical. (Of course, purifications themselves must be validated.) Development chemists will recognize this situation: If one tries to optimize a reaction, one must not crystallize the product because the purification will hide the variations the optimization is trying to detect. This is not so with formulations leading to drug products, for which "what goes in comes out" and for which the process, therefore, defines the quality of the product. For these two reasons, process validation, which, by defining the limits of the process, "builds quality in up front of the process," is less critical for drug substances than for drug products.

At the same time, validation is more difficult for drug substances than for drug product, precisely because purifications remove variability and because each chemical reaction generates impurities. In other words, what goes in does not come out in a synthesis, as it does in a formulation. On the contrary, the amount of impurities increases during the reaction, then decreases during purification, so the quality of the starting material is not necessarily related to the quality of the product; building in quality up front is more difficult for drug substances, because it is more difficult to validate such a fluctuating system. Last, chemical reactions contain multiple and interdependent reaction variables; we will see in a moment the consequences of this fact and how to deal with it in process validation.

To summarize, using Fig. 8.2, I will again define process validation and illustrate why process validation is more difficult for synthetic processes than for dosage formulations.

Let us choose one reaction variable, e.g., reaction temperature, and see how it affects the stereoselectivity of the reaction (Fig. 8.2). During process optimization, the process chemist would typically vary the reaction temperature in several experiments and look at the resulting stereoselectivity; in this way he or she would

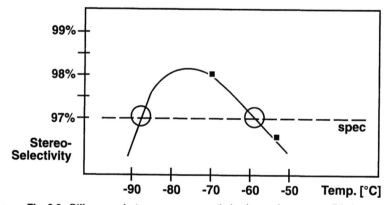

Fig. 8.2. Differences between process optimization and process validation.

quickly climb the curve in the diagram toward the optimum. Using chemical intuition and educated guesses, only a few experiments are needed to get near the maximum. Process validation, on the other hand, is the opposite activity. Process validation, by the narrow definition, means finding the unique points on the curve (indicated by circles in Fig. 8.2) at which the process fails specifications, set at 97% in this example and indicated by the horizontal line in the figure. Now the chemist is sliding down the curve, away from the optimum; he or she is taking the process apart and stressing the process until its edges are found. The two activities are thus opposing, but good process development would contain both process validation and process optimization, as such information is valuable to the plants.

Process validation is thus defined as "finding the limits of process variables" and is only a part of process development. Process development consists of process optimization plus process validation plus scale ups plus proving reproducibility, and so on (Fig. 8.3).

These activities may overlap, as, for example, by chance one may hit the maximum during validation or the edge during optimization, but the goals are quite different for the two activities.

8.4.3. More Experiments

At the beginning of this chapter, I said that more experiments were needed in the area of process validation than was usual for process development (6). There are two reasons: first, to find a unique point on a steep curve requires more trial and error (similar to regression analysis) as opposed to the more usual optimization that requires a process chemist to get to the flat part of the curve where there is a wider range of acceptable results (e.g., in Fig. 8.2, anything between -60 and $-85°C$ would pass specifications). A process chemist is used to working near the optimum and not at the edge of the process where half the experiments will fail specifications. Nevertheless, as we will see later, it is valuable to the production plant to know where the edges are. Second, and more serious, chemical reactions have multiple and often interdependent variables. Trying to define limits of parameters for all permutations thus means not only defining a curve but rather constructing a multidimensional surface, which is not practicable to do in chemistry, because each point on this surface requires an experiment. We will see in a moment how to simplify this task, however.

Although computer programs exist to plot such graphs and to predict maxima, statistical analysis is not useful for chemical development, for several reasons. Such

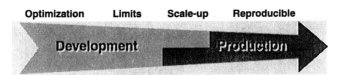

Fig. 8.3. Components of process development.

a program requires lots of data points and may be useful for production (with numerous batches), whereas in process development we have only a few data points at the beginning. More critical, chemistry is discontinuous. We are not dealing with a straight line, not even with a symmetrical parabola. For example, if one cools the reaction described in Fig. 8.2 further, at $-90°C$ the starting material suddenly precipitates, and the stereoselectivity drops dramatically. Similarly, if one heats it too high, the product may suddenly decompose; statistical analysis cannot predict such breaks in the curve. In other words, chemistry is not mathematically precise, it is still an experimental science, and while mathematics can calculate maxima, it is not good at predicting the edges of the process, which by definition is the discontinuity or, at best, an artificial point of crossing the specification.

8.4.4. Types of Ranges

Several times in this chapter, ranges of variables were mentioned. There are several types of ranges, which is important to keep in mind whenever we talk about defining limits of reaction parameters (Fig. 8.4). The narrowest range 1, is the master procedure range, the small range prescribed in processes over which the plant will fluctuate because of inherent equipment limitations and momentum, e.g., the temperature can be kept only within $3°C$ of any setting. The next wider range, 2, would be the verified range, i.e., the range of values actually experienced in the plant, either because of intentional changes or because of deviations. Such experience ranges can be used for retrospective validation (as one then knows whether the product made in that range will pass or fail). The third range, 3, is wider yet and would represent the practical achievable limits. For example, if one is in a steam-heated reactor, the practical limit to the temperature would be $150°C$ and no higher. Finally, there is range 4, where the process fails specifications, for whatever reason. It is dangerous and expensive to fail processes on a plant scale, so the plant will always try to stay safely away from the edges, in ranges 1 or 2, whereas the laboratories (where failure is not critical) may want to perform prospective validation in ranges 3 or 4. The diagram is intentionally drawn as a telescope that one can compress, because sometimes range 4 is already inside range 3 or even 2, but then those experiments will catch the limits of parameters.

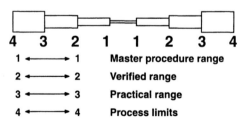

Fig. 8.4. Types of ranges for process variables.

8.4.5. Guide to Drug Substance Validation

These are the steps in process validation (defined narrowly as the finding of process limits).

1. *Choose the final process.* Obviously one should validate only the final process, as any major changes would require repeating the validation. One cannot validate a changing process, but one validates it by testing changes; this statement is one of those paradoxes.

2. *Choose your validation protocol.*

3. *Identify critical parameters in final steps of the synthesis.* By final steps we generally mean, for example, from the point in the synthesis at which the carbon skeleton of the molecule is put together, or from the point at which stereochemistry is defined, or from the point at which critical purification occurs (e.g., separation of isomers), and so forth. One does not have to validate the entire synthesis.

4. *Scale down the process.* Prospective validation is usually performed in the laboratory as one does not want to stress the process on a plant scale, it is too dangerous and expensive, not to mention ecologically irresponsible. However, scale is also a critical variable; this fact is a weakness of performing validation on laboratory scale. It is also an example of the Heisenberg uncertainty principle (7), which says: "whenever one measures a system, one disturbs it, therefore, one can never measure a system." The same is true with process validation: we are trying to build in reproducibility by making changes. Nevertheless, process validation is like building guard rails on a highway to ensure that the production plant will always stay on the road.

5. *Vary one parameter at a time over a practical range.* This limitation is a compromise to avoid the infinite number of experiments trying to define the multidimensional surface for the reaction, as mentioned earlier. One varies only *one* parameter at a time and only over a *practical* range. This means that even if the process edges are not encountered inside the practical range, we do not look for them out further, because by definition, the plant cannot exceed the practical range anyway.

8.4.6. Example of Process Validation

Critical parameters are usually chosen based on the chemist's experience with the reaction and on his or her learned intuition. Experiments during process validation may prove that a chosen variable is not critical after all (if changing it does not alter the outcome of the reaction), or unexpected variables may be discovered to be critical. In this example, the following were judged to be critical parameters: amount of reagent, reaction temperature, number of washes, and purity of crude product.

Each critical parameter is then varied between a low, standard, and high value. Table 8.7 illustrates the experiments varying the amount of a reagent at 0.9 eq (low value), 1.0 eq (standard process value), and 1.1 eq (high value). Again, this table is only an excerpt; the actual table would have many more lines (but not columns). We see that for this variable, an excess of the reagent (1.1 eq) is detrimental: the

Table 8.7. Process Validation—Amount of Reagent

Product	Amount of Reagent		
	0.9 equiv	1.0 equiv	1.1 equiv
Yield, %	73.7	75.3	40.1[a]
Purity, %	97.0	97.3	89.5[a]
Unknowns, %	0.4	0.1	3.0

[a]Fails.

product fails all specifications, so excess must be avoided. Such information is valuable to the plant, and this example shows why process validation is useful. Similarly, the reaction temperature was varied between low, standard, and high values (Table 8.8).

At the high end, the purity of the product fails specification, leading us to conclude the plant must avoid heating this reaction above 30°C. Here we encounter the different types of ranges: $25 \rightarrow 30$°C is range 1, the plant procedure range, whereas range $15 \rightarrow 40$°C is the verified range 2 (we have experience over this range); however, we have already reached the edge, so it is actually range 4 (compress the telescope in the Fig. 8.4).

The process chemist felt that the number of extractions was a critical parameter, and he was right (Table 8.9). After only one wash, the amount of unknown by-products failed specifications; the conclusion is then that at least two washes are needed, but at five, the quality starts to go down again.

An important parameter to determine is the effectiveness of any purification method. Table 8.10 follows a purification method step by step to see what is accomplished with each operation. The crude product (97.6% pure) contains large amounts of by-products; slurrying (stirring with solvent without dissolving the solid) removes mainly by-product 2 and most of the unknowns, and the crystallization removes most of the rest of the by-products to give an excellent purity of the drug substance. However, the success of the crystallization depends on the purity of the crude, so the latter is a critical variable. Table 8.11 proves the point.·

If the crude product is only 93.5% pure, for example, because it contains 6.5% of by-product 1, one crystallization will not remove all of it, and the product will fail specifications. It is based on such experiments that specifications for intermediates are set; here we would set the specification for by-product 1 in the crude precursor at <3%, because the success of the purification with this quantity was validated.

Table 8.8. Process Validation—Reaction Temperature

Product	Temperature, °C		
	15–20	25–30	35–40
Yield, %	73.0	75.3	73.3
Purity, %	97.1	97.3	93.2[a]
Unknowns, %	0.1	0.1	0.2

[a]Fails.

Table 8.9. Process Validation—Number of Extractions

Product	Number of Washes		
	1	2	5
Yield, %	76.4	75.3	72.2
Purity, %	97.6	97.3	96.0
Unknowns, %	1.2[a]	0.1	0.7

[a]Fails.

Table 8.10. Process Validation—Purification Method

Analysis	Purification		
	Crude →	Slurry →	Crystallization
Product, %	97.6	98.0	99.9
By-product 1, %	2.4	2.0	0.1
By-product 2, %	1.2	0.3	0.1
Unknowns, %	0.7	0.1	0.0

Table 8.11. Process Validation—Crude Product Purity

Analysis	Purification		
	Crude →	Slurry →	Crystallization
Product, %	93.5	94.6	97.5
By-product 1, %	6.5	5.4	2.5[a]
By-product 2, %	1.2	0.1	0.1
Unknowns, %	0.8	0.1	0.1

[a]Fails.

All these experiments are described in the development report, and this brings us back to the beginning of this chapter, which predicted that the FDA guidelines mean a lot more documentation and a little more experimentation for chemical development.

BIBLIOGRAPHY

1. *Code of Federal Regulations,* Title 21, Parts 210 and 211, 1978.
2. *Guide to Inspection of Bulk Pharmaceutical Chemicals,* U.S. Food and Drug Administration, Rockville, Md., 1991.
3. S. W. Blum, "Drug Impurity Testing: Issues and Approaches," paper presented at the Ad Hoc FDA Advisory Committee Meeting, June 22, 1993.

4. C. J. Ganley, "Impurities from a Clinical Perspective," paper presented at the Ad Hoc FDA Advisory Committee Meeting, June 22, 1993.

5. *Guideline on General Principles of Process Validation,* U.S. Food and Drug Administration, Rockville, Md., 1987.

6. R. Carlson, *Design and Optimization in Organic Synthesis,* Elsevier, Amsterdam, 1992.

7. G. W. Castellan, *Physical Chemistry,* Addison-Wesley, Reading, Mass., 1964, p. 413.

9

THE FUTURE

One often hears that organic chemistry is a "mature science." Does this mean it is over the hill? I do not think so. The term *organic chemist* implies an intimate knowledge of carbon-containing molecules and their reactions, and so the organic chemist may be among the best equipped to solve the mysteries of biochemistry, to discover the mechanisms of diseases, and to design cures for these diseases on the molecular level. Granted, these cures will require new types of therapies (e.g., enzymes, antibodies, or genes), but these are best made by organic chemists also. The field is better called biochemistry, not molecular biology.

I believe that in the future organic synthesis will be supplemented but not displaced by fermentation, genetic engineering, and mammalian-cell growth. The small organic molecules that will still be needed, e.g., peptidomimetics, will be synthesized with increasingly sophisticated organic chemistry methods.

A most exciting new method is the use of monoclonal antibodies as catalysts in organic synthesis. Catalytic processes are ecologically sound because, by definition, they employ only small amounts of recyclable reagents, namely catalysts. The desirable qualities of a catalyst are that they accelerate the reaction rate, they have high (regio- and enantio-) specificity, and they are biodegradable.

Natural enzymes meet these requirements; their limitation is, of course, that we are restricted to those enzymes that had been isolated from nature, and furthermore, that the scope of reactions they can catalyze is limited.

One can custom-make enzymes by genetic engineering, but this is laborious. A better way is to let living organisms not only manufacture such catalysts but also custom-design them for a specific reaction or application; this feat can be accomplished with monoclonal antibodies (MAbs). MAbs as catalysts in organic synthesis are a revolutionary new method, especially for process development chemists, as it allows for the unprecedented custom design of catalysts for specific reactions.

By lowering the energy of the reaction transition state, a reaction is accelerated. This can be accomplished by preparing a MAb that binds to the transition state and lowers its energy. Such an antibody can be elicited (in mice) with a stable hapten resembling the transition state configuration of a given reaction. Note that the hapten, i.e., the transition state analog, is usually coupled with a carrier protein to form an antigen. For this purpose, a carboxylic acid side chain is often attached to the hapten.

For example (these examples are research proposals that have not been tried yet), the hydrolysis of ester **1** (Fig. 9.1), which goes through a tetrahedral transition state **3** (Fig. 9.2), can be made enantiospecific by adding an antibody that will bind to only one enantiomeric transition state.

A phosphonate is usually used as the hapten because it is a stable functional group that resembles the tetrahedral transition state of a saponification. Thus an antibody to enantiopure hapten **4** would bind to only one enantiomer of the transition state **3** in the ester hydrolysis; it would presumably lower its energy and in turn ac-

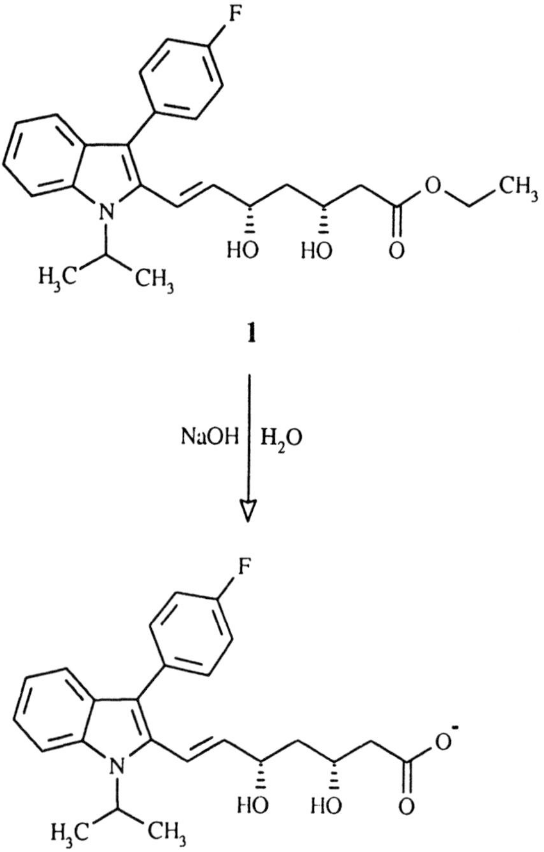

Fig. 9.1. Saponification of an ester.

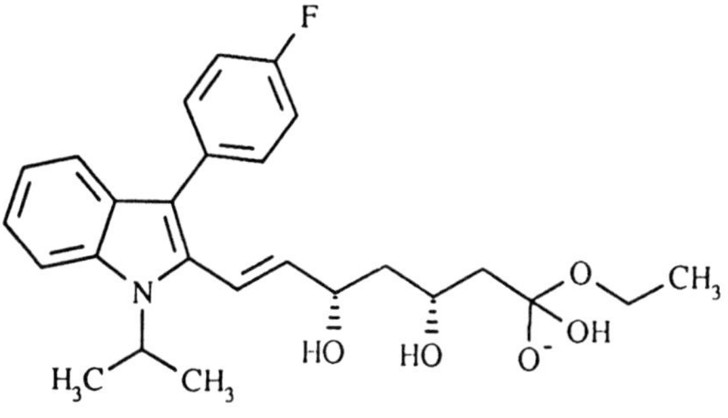

3

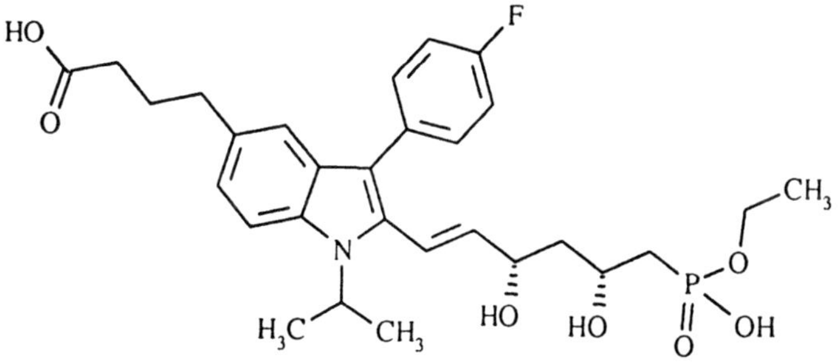

4

Fig. 9.2. Transition state of a saponification and its stable analog.

celerate the reaction, resulting in a catalytic enantioselective hydrolysis to give enantiopure **2**.

By this method, even unnatural reactions can be catalyzed. For example, a Cope rearrangement of **5** (Fig. 9.3) goes through a tricyclic transition state **6** (Fig. 9.4).

A stable hapten, mimicking the transition state, could be the tricycle **7**. An antibody raised by this hapten would accelerate this reaction and allow it to proceed at a lower temperature, for example.

Similarly, the Diels-Alder reaction in Figure 9.5 climbs over a bicyclic transition state **8** (Fig. 9.6), which loses acetylene to give the desired product **9**. This two-step Diels-Alder reaction makes the design of the transition state analog easier, as it avoids the product-inhibition problem (if the transition state resembles the product too much, the action of the antibody is inhibited by the formation of the reaction product, to which it would bind). A stable transition-state hapten could be the carbon analog **10** (Fig. 9.6).

Another thermal or photochemical cyclization (Fig. 9.7) leading to indoline **11**, proceeds via ylid **12** as its transition state (Fig. 9.8). It is amusing that the carbon can be replaced with a nitrogen and the nitrogen with a carbon to form a stable transition state analog and hapten **13** (Fig. 9.8).

5

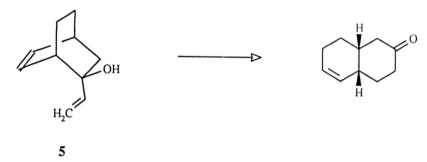

Fig. 9.3. Cope rearrangement.

6

7

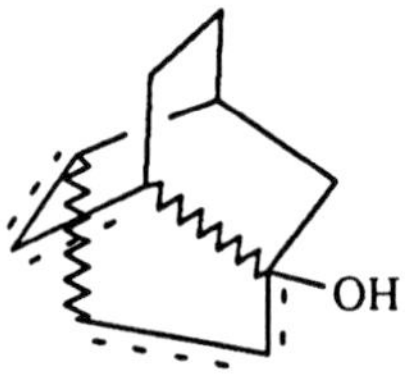

Fig. 9.4. Transition state of a Cope rearrangement and its stable analog.

9

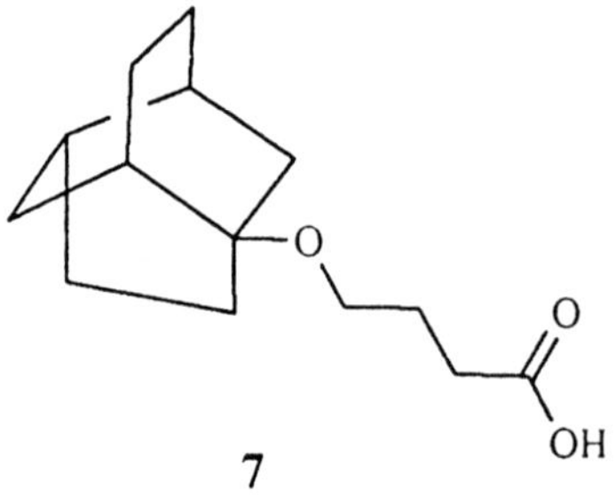

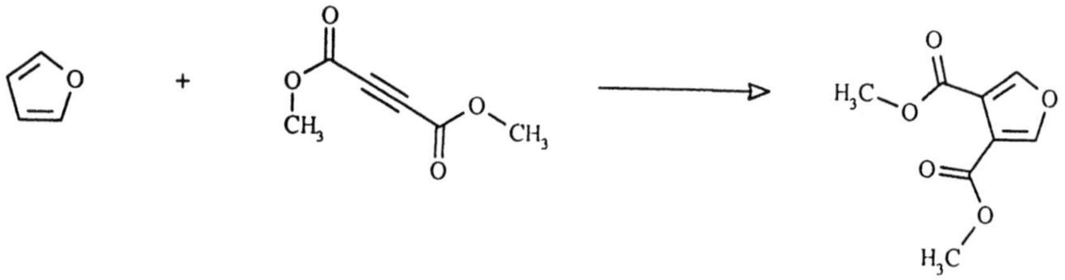

Fig. 9.5. Diels–Alder reaction.

8

10

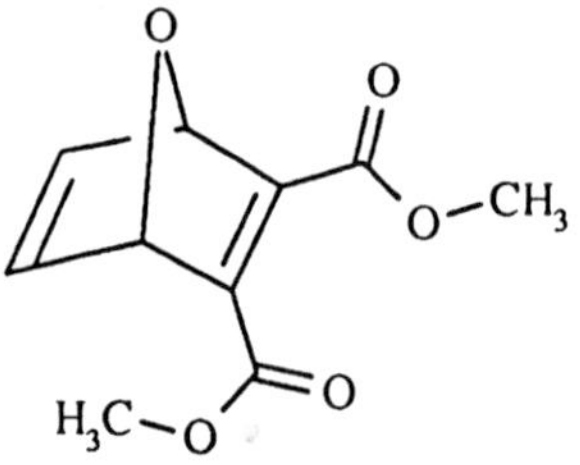

Fig. 9.6. Transition state of a Diels–Alder reaction and its stable analog.

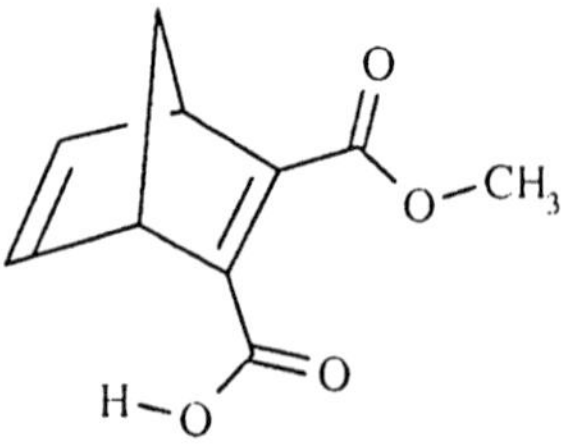

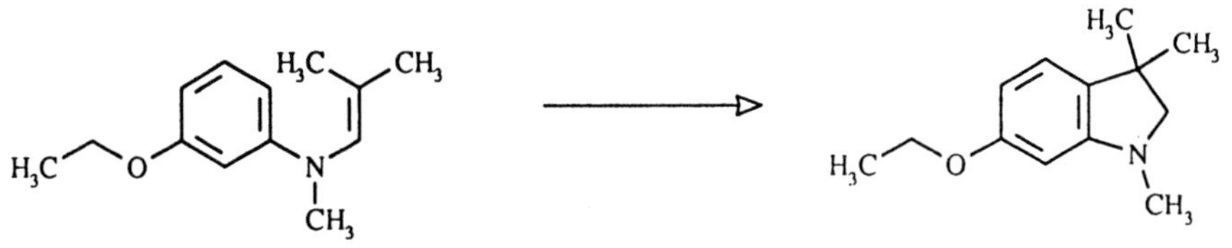

Fig. 9.7. Thermal or photochemical cyclization.

Finally, in a nucleophilic aromatic substitution reaction, the addition of ammonia to 2-bromopyridine **14** (Fig. 9.9) goes through a transition state **15** to give 2-aminopyridine **16**. The heteroatoms can be replaced with oxygen and carbon, to synthesize a stable hapten **17** (Fig. 9.10). An antibody raised by this hapten would catalyze the substitution **14 → 16** and accelerate it.

There are other strategies that can be used with MAbs, not just stabilization of transition states. For example, more advanced methods would (*1*) add catalytic sites in an appropriate position in the MAb by eliciting them with hapten complementarity or by site-specific mutagenesis or semisynthetically, (*2*) add a cofactor binding site, and (*3*) lower the entropy of a reaction by bringing two reactants together (Fig. 9.5 is an example). Antibodies can be used for various applications: to study structure–activity relationships (SAR), to splice antibody genes into bacteria or mammals (to catalyze desired unnatural reactions, either to prepare products or to allow the growth of these organisms on unnatural nutrients, e.g., environmental pollutants), to design sequence-specific proteases or peptidases, to design enantioselective redox reagents, or to be used directly as therapeutic agents (by incorporating therapeutic agents proximate to a MAb combining-site, the MAb will deliver the agent to the appropriate site or cell). This field, and indeed organic synthesis, is limited only by our imagination.

The same holds true for science in general. Many doomsayers (but responsible people who make policies) blame science for our environmental or even health problems; many consider scientists self-serving, self-contained, working with entirely circular arguments. However, unlike philosophy of the antique, unlike theology of the Middle Ages, science of the twentieth century holds the truth. Its proof, of course, is in the experimental evidence. How could we synthesize the cancer

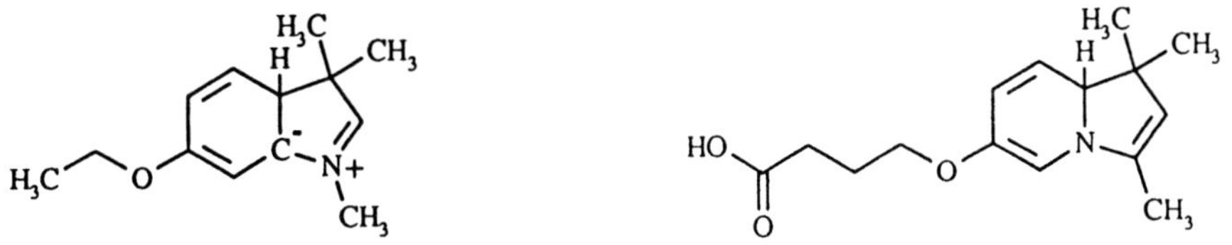

Fig. 9.8. Transition state of an electrocyclic reaction and its stable analog.

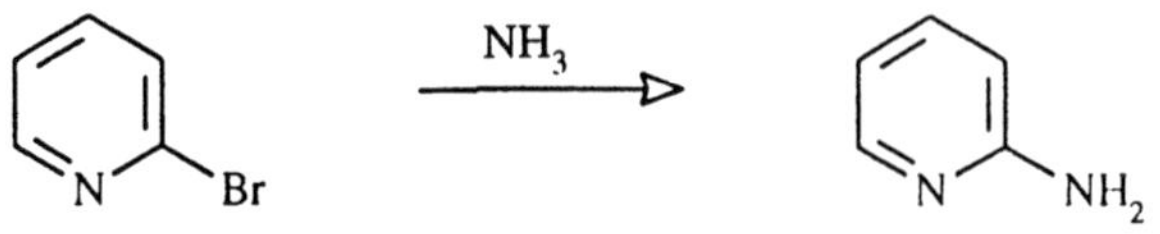

14 **16**

Fig. 9.9. Nucleophilic aromatic substitution.

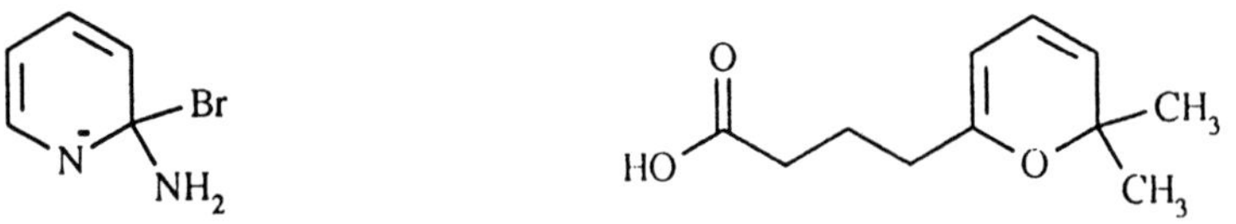

15 **17**

Fig. 9.10. Transition state of an SₙAr reaction and its stable analog.

drug paclitaxel, how could we make a 200-t airplane fly, or how could we land a robot on Mars unless our scientific theories were true? Furthermore, science is amoral. Science simply consists of knowledge; if it is abused for immoral purposes, one must blame the perpetrator not the tool (knowledge in this case). Admittedly, knowledge is a fierce weapon, but when used wisely it can improve and, indeed, save human lives, as do pharmaceuticals.

SUGGESTED READINGS

I. Fujii, R. A. Lerner, and K. D. Janda, "Enantiofacial Protonation by Catalytic Antibodies," *J. Am. Chem. Soc.*, **113**, 8528–8529 (1991).

J. Hasserodt, K. D. Janda, and R. A. Lerner, "Antibody Catalyzed Terpenoid Cyclization," *J. Am. Chem. Soc.*, **118**, 11654–11655 (1996).

D. Hilvert, K. W. Hill, K. D. Nared, and M.-Teresa M. Auditor, "Antibody Catalysis of a Diels-Alder Reaction," *J. Am. Chem. Soc.*, **111**, 9261–9262 (1989).

S. Ikeda, M. I. Weinhouse, K. D. Janda, and R. A. Lerner, "Asymmetric Induction via a Catalytic Antibody," *J. Am. Chem. Soc.*, **113**, 7763–7764 (1991).

K. D. Janda and co-workers, "Chemical Selection for Catalysis in Combinatorial Antibody Libraries," *Science*, **275**, 945–948 (1997).

T. Kitazume, J. T. Lin, T. Yamamoto, and T. Yamazaki, "Antibody-Catalyzed Double Stereoselection in Fluorinated Materials," *J. Am. Chem. Soc.*, **113**, 8573–8575.

S. J. Pollack and P. G. Schultz, "A Semisynthetic Catalytic Antibody," *J. Am. Chem. Soc.*, **111**, 1929–1931 (1989).

T. S. Scanlan, J. R. Prudent, and P. G. Schultz, "Antibody-Catalyzed Hydrolysis of Phosphate Monoesters," *J. Am. Chem. Soc.*, **113**, 9397–9398 (1991).

P. G. Schultz, R. A. Lerner, and S. J. Benkovic, "Special Report on Catalytic Antibodies," *Chem. Eng. News*, May 28, 1990, pp. 26–40.

D. Seebach, "Organic Synthesis—Where Now?" *Angew. Chem. Int. Ed. Engl.* **29,** 1320–1367 (1990).

H. Suga, N. Tanimoto, A. J. Sinskey, and S. Masamune, "Glycosidase Antibodies Induced to a Half-Chair Transition-State Analog," *J. Am. Chem. Soc.,* **116,** 11197 (1994).

T. Uno, J. Ku, J. R. Prudent, A. Huang, and P. G. Schultz, "An Antibody Catalyzed Isomerization Reaction," *J. Am. Chem. Soc.,* **118,** 3811–3817 (1996).

INDEX